WHEN DISASTER STRIKES

EXTREME FIRES AND FLOODS

Written by John Farndon

外语教学与研究出版社
FOREIGN LANGUAGE TEACHING AND RESEARCH PRESS
北京 BEIJING

In just a few seconds, a spark can turn a forest into an inferno, igniting hundreds more fires – and then the fight is on to control the rampaging flames. Follow the scorched trail of the worst wildfires ever, with firefighters battling them sometimes for months on end...

Water, the enemy of fire, is vital to life – except in the deluging amounts brought by storms and rising seas. See how extreme floods reduce land to swamp, sweep away houses and bridges, and bring hunger and disease. And discover how technology helps to predict them – it's now more important than ever.

CONTENT

WILDFIRES

Wildfires are uncontrolled fires that sweep across field and forest. They can burn for days or weeks, wiping out all life over vast areas, and threatening human settlements close by. They may be ignited by the glow from a discarded cigarette or a stray ember from a barbecue, which turns an entire forest into a raging inferno, consuming everything in its path.

FLOODS

A flood happens when a lot of water suddenly submerges normally dry land. It can rise and recede quickly in a 'flash' flood after an intense storm, or build up slowly and last for months. Severe floods can be devastating, swamping vast regions and destroying people's homes and livelihoods, often claiming many lives.

HOW DOES A WILDFIRE HAPPEN?

Wildfires start when some kind of spark ignites dry vegetation. They occur wherever there is enough moisture to let trees and bushes grow, but where there are long warm, dry summers. As the heat of summer dries them out, the trees and bushes turn into a giant pile of firewood and kindling – just waiting for that dangerous spark to set them alight.

UNBELIEVABLE!

In 2012, nearly 40,469 km^2 of forest burned in the USA – an area bigger than these three US states combined: New Jersey, Connecticut and Delaware. On average, twice as large an area of forest burns in the USA now as it did 40 years ago.

CROWN OF FIRE

Wildfires can burn in different ways. Ground fires burn slowly in the soil, consuming the organic material there. Surface fires spread rapidly along the ground burning fallen leaves and branches. Crown fires are the most dramatic of all, burning ferociously with huge flames that spread from treetop to treetop, catching all alight.

FIERY TRIANGLE

To burn, a fire needs three things, sometimes called the fire triangle: fuel, oxygen and heat. The fuel is provided by vegetation. Oxygen comes from the air – when fuel burns, it reacts with the oxygen. Heat comes from the spark that starts the fire, or from the fire itself, once it is going. The fire will only stop when deprived of one of these three.

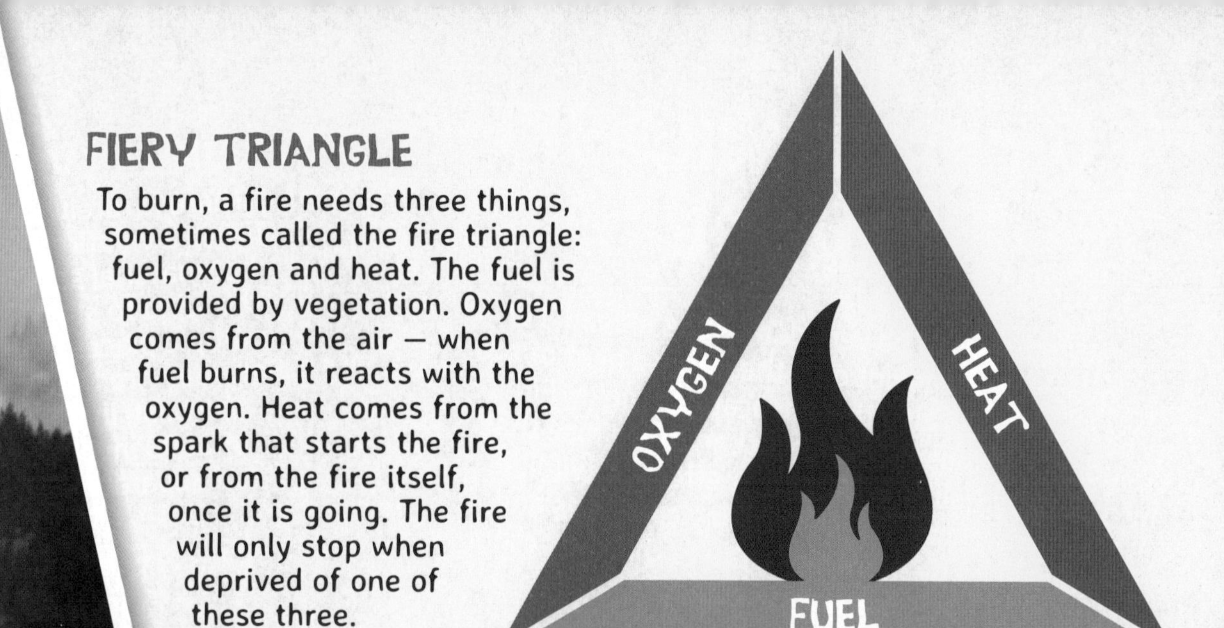

DANGEROUS SPARKS

Fires can be started in many ways. They can start naturally when lightning from a summer thunderstorm sets vegetation alight, or with hot cinders from a volcanic eruption. But many fires are ignited by people, either carelessly with a dropped match or a camp fire, or deliberately by arsonists to cause mayhem.

FIRE WHIRLS

Wildfires, especially big crown fires, create their own winds, as they draw air in rapidly. Sometimes these winds, although local, can be as fierce as a hurricane. In certain conditions, tornadoes called 'fire whirls' can start within a wildfire, as air roars up in a spiral, carrying flames with it.

EXTREME WILDFIRE DAMAGE

Wildfires can be very dangerous indeed. For people, there is not only the intense heat of the flames, but also the suffocating smoke. Fires destroy houses and almost anything else that gets in their way. And their effect on wildlife, especially on forest-dwelling creatures and plants, can be more devastating than any other natural disaster.

UNBELIEVABLE!

There is rarely just one fire in a wildfire, because hot material spreads on the wind to start fires in multiple places. The Black Saturday fire that engulfed the bush in Australia in 2009 involved 400 separate fires.

BURNED OUT

Wildfires spread fast and can quickly change direction. People fleeing by car may misjudge which way a fire is spreading, get lost in the smoke or find their route blocked by flames. That's why it is vital to listen to warnings from the authorities and get out of the area early if you're ever close to a wildfire.

REBIRTH

For some plants and animals, a fire is not a disaster but an opportunity. The fire clears away dead wood, and makes room for new plants. Many forests bounce back after a fire. Jack pines, white pines and yellow birch have their seeds opened up and given a nutritious new ash-rich soil.

ANIMAL TRAGEDY

For many forest creatures, a wildfire is a disaster. Birds may fly away. Bigger mammals may run. Some creatures may burrow into the ground, or take cover under rock or in water. But young and small animals face certain death, especially tree-dwelling animals like squirrels.

BIG BURN

In 1910, fire engulfed 12,141 km^2 in Idaho, Montana and Washington in the USA and killed at least 85 people. Whole towns were burned away, and hundreds of soldiers were brought in to try and stop the flames. From then on, the Forest Service determined to put out every fire. But now many fire experts think this is not only dangerous for firefighters but does little to save forests.

FIGHTING AN EXTREME WILDFIRE

Wildfires very quickly get out of control and burn over such a vast area that it is hard to fight them – and very dangerous to do so. Different fires need different approaches, but the basic idea is always the same – to rob the fire of its fuel, so that it eventually burns out by itself.

EYEWITNESS

Firefighter Drew Miller stops fires by burning away their fuel source, and uses "chainsaws... to remove fuel sources and dig lines so the fire can't spread as easily. Then we clear the area behind our friendly fire."

FIRE MAP

Computer and satellite technology has enabled experts to keep an ever closer eye on wildfires in the USA. They can now show live, interactive maps of just where a wildfire is burning at any one time. The high concentration of fires in the forest regions of the Northwest becomes very clear this way.

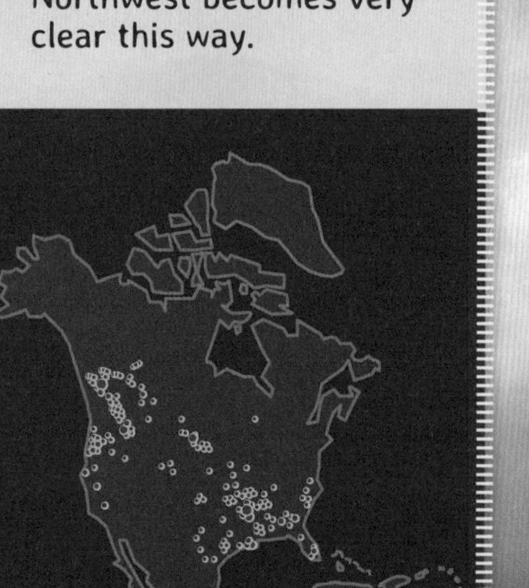

CHOPPER DROP

Special helicopters are sometimes flown over the fire to dump water, fire retardant and other chemicals, such as ammonium phosphate, that slow down fires. They may water bomb fires with as much as 7,000 litres at a time. They are not trying to put out the fire, though – just buying the ground crew time to create a break in the wood to stop the fire.

WATCHING FOR SMOKE

In forest areas prone to wildfires, watchtowers are often built so that fire-watchers can keep an eye on every potential fire. They might track and record a lightning strike, then keep scanning the horizon for plumes of smoke.

EARLY WARNING

Satellite technology means people are much less likely to be caught by surprise than in the past. Sensitive satellites monitor slight changes in the colour tone of the forest that might indicate smoke. They can then feed this into national wildfire monitoring systems to issue people nearby with a warning.

FIREFIGHTERS

Fighting a fire on the ground is a very dangerous job, and requires a high degree of training. Firefighters are given special protective clothing, such as shirts and trousers made of Nomex, and oxygen supplies to help them when they breathe in the smoke.

EXTREME WILDFIRE STORY

Wildfire disasters can strike out of the blue. On 1 May 2016, the town of Fort McMurray in Alberta, Canada, suddenly found itself facing utter destruction as a wildfire raged in the forests beyond. Miraculously, only two people died, yet 2,400 houses were destroyed and the disaster was one of the costliest in Canadian history. The insurable losses were about $2.75 billion.

EYEWITNESS

Fire officer Bendfeld: "I was watching a fire that was going across the trees... it was moving at 28 metres per hour. Then it was going at 32 metres per hour, then 300 metres per hour... How do you get a hold of that?"

EYEWITNESS

Text messages between Jody Butz out firefighting and his 15-year-old son Wesley stuck at school, realising they might lose their home:

Butz: *Lock down the school and stay with the boys. Be strong.*

Wesley: *Is the fire east of the golf course... towards the house?*

(Butz knew it was, but couldn't be sure where it would end up.)

Butz: *It's coming. Be brave.*

FIRE HELL

Fires are likely in dense, dry forest in hot weather. But the wind made the Fort McMurray fire especially frightening and hard for firefighters. High winds scattered hot embers, starting new fires in unexpected places. They also whipped up flames, so a fire eating slowly forwards suddenly flared up and roared through the trees.

GET OUT OF THERE!

At first, the fire seemed to be under control. Then on the afternoon of 3 May, the wind changed direction and the fire, nicknamed 'The Beast', roared back, heading straight for the town. The government ordered everyone of Fort McMurray's 80,000 inhabitants to leave at once and placed some of the evacuees in lodge camps.

ROADBLOCK

With the fire forever changing direction, the authorities were not sure which way to send people trying to flee the fire. The police set road blocks to stop them heading into even more danger. Vast traffic jams built up at the blocks. Many people stuck there were frightened and angry, sitting on the highway while their homes and livelihoods burned.

FIRE FROM ABOVE!

A satellite view shows the huge plumes of smoke billowing from the Fort McMurray fire. By mid-May the fire had spread into neighbouring Saskatchewan. By late June, rain, cooler weather and the extraordinary work of the firefighters had brought the fire under control. But it wasn't completely out until 2 August,2017.

COASTAL FLOODING

Low-lying coasts are very vulnerable to flooding from the sea. Sea water may be pushed on to land by strong winds and hurricanes. And the flooding may be even worse if the winds are carrying heavy rain. Some of the most devastating coastal floods, though, come from tsunamis – vast waves set in motion by earthquakes under the sea.

UNBELIEVABLE!

The highest ever waters in the USA occurred during Hurricane Katrina. A 6.7 m storm surge, with waves over 3.5 m on top, built waters up to an incredible 10.2 m at the Beau Rivage Lighthouse in Biloxi, Mississippi.

KATRINA FLOOD

When Hurricane Katrina struck New Orleans in August 2005, it wasn't the winds that did the real damage but the flooding that came a day later. The waterways running through the city were filled to bursting by the combination of the storm surge (opposite) and the heavy rain brought by the storm. Finally, the levees, the barriers containing the water, burst and the water flooded into the city.

STORM SURGE

Powerful storms, especially hurricanes, can cause a sudden and dramatic rise in the sea, called a storm surge. It's like a super-high tide. Low air pressure in the storm's centre makes the sea swell upwards, and high winds push water towards the shore and build up waves. Storm surges can briefly raise the sea level many metres, swamping coastal areas.

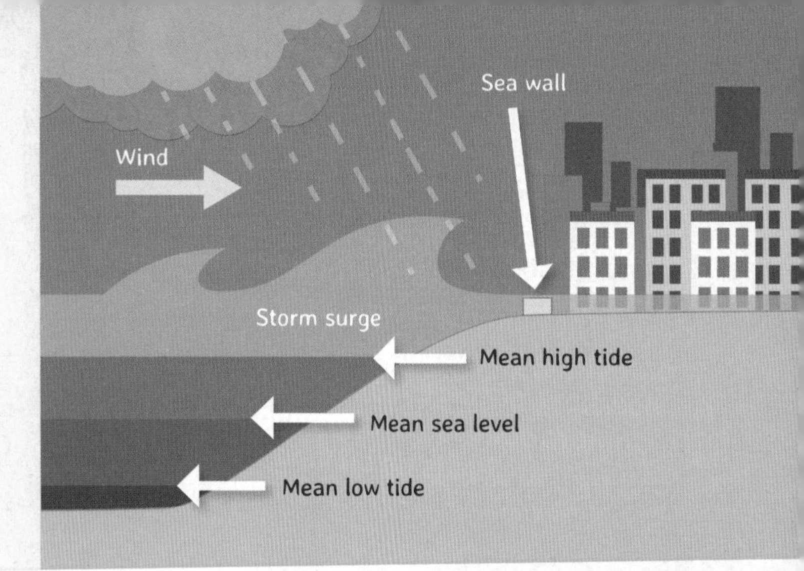

STORM WAVES

Even without a storm surge, storms can bring monster waves that smash against sea defences and may breach them. But when these waves coincide with a high tide or a storm surge, they can bring devastating floods.

HIGH WATER

The height of sea tides varies as the Moon makes its month-long journey round the Earth. They are at their most extreme during 'spring tides' when, twice a month, the Moon is in line with the Sun. Then, the gravitational pull of the Moon and Sun combines to raise and lower the sea dramatically. In between come more moderate 'neap tides', when the Moon and Sun pull at right angles.

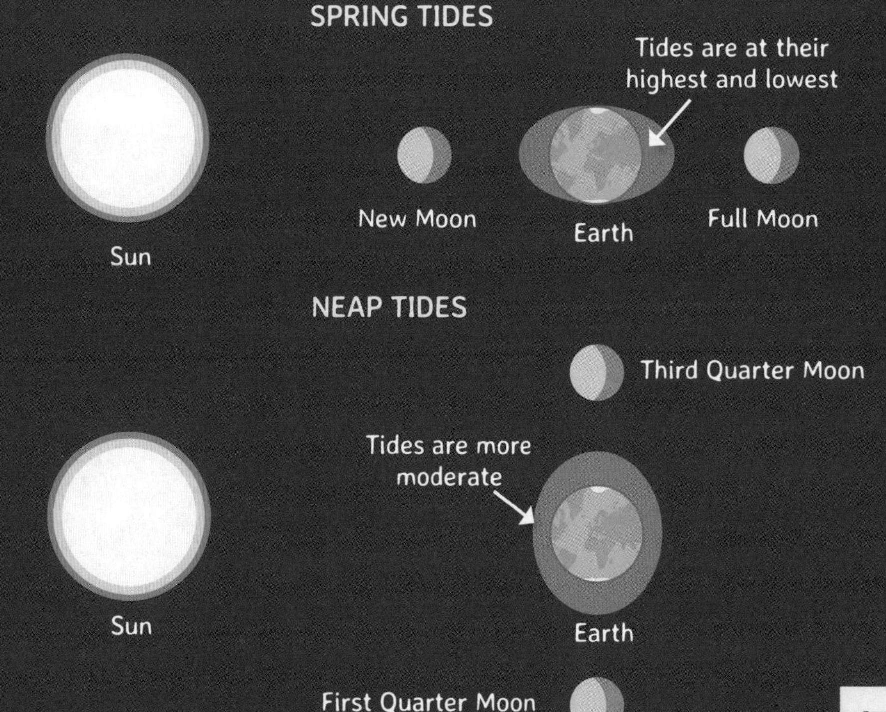

RIVER FLOODING

Floods happen when there is too much water to run away safely through drains and rivers. They almost always follow heavy rain or melting snow. Then rivers or lakes may overflow, or there may be too much rain for it to soak into the ground and so it builds up on land.

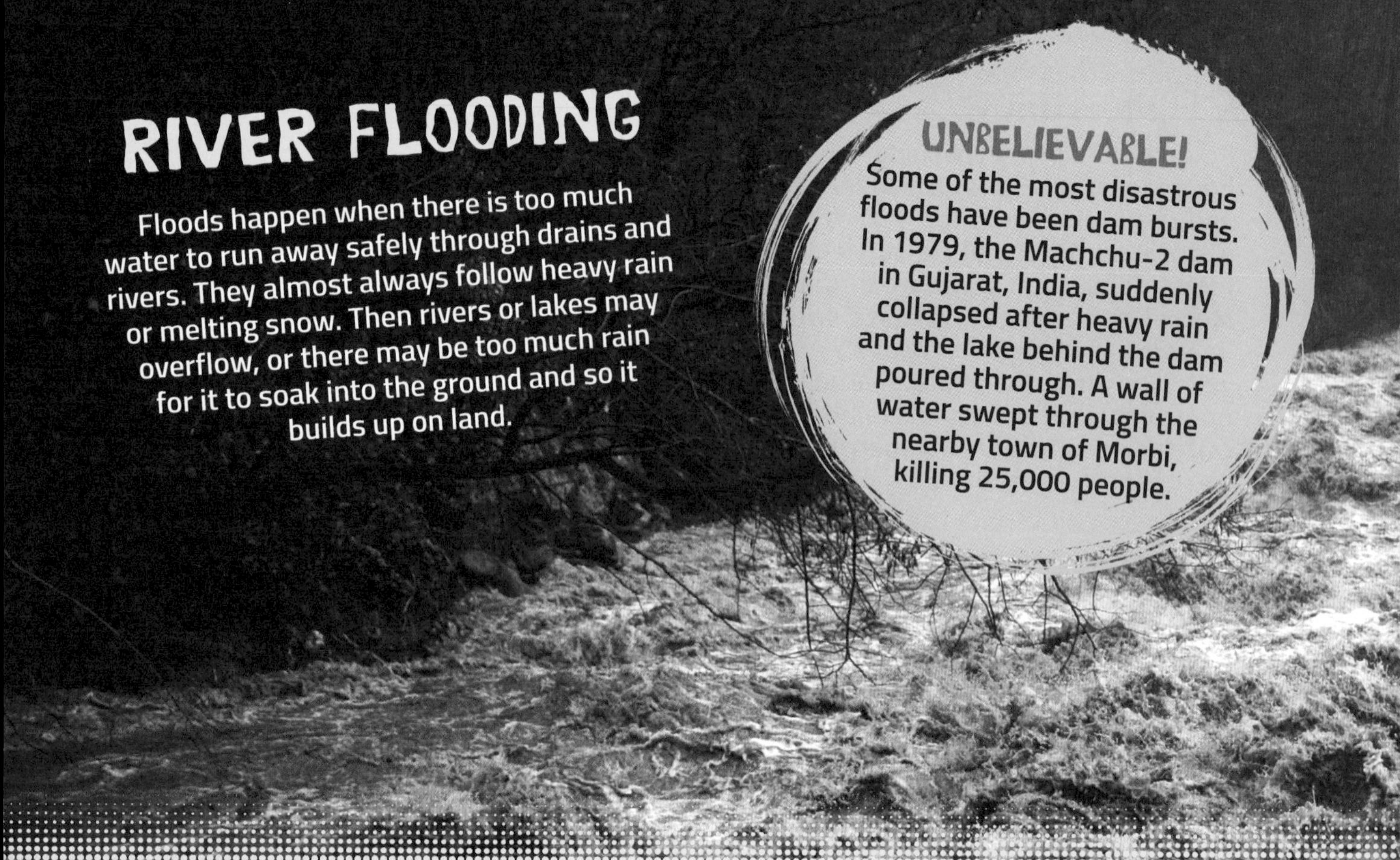

UNBELIEVABLE!

Some of the most disastrous floods have been dam bursts. In 1979, the Machchu-2 dam in Gujarat, India, suddenly collapsed after heavy rain and the lake behind the dam poured through. A wall of water swept through the nearby town of Morbi, killing 25,000 people.

FLASH FLOODS

Most floods develop slowly, but 'flash floods' occur in just a few hours or even minutes! They start with rain so heavy it cannot soak into the ground fast enough. Instead, it runs straight off in a surge powerful enough to wash away cars, bridges and buildings! Most flash floods occur after sudden storms in dry areas. Sometimes, flash floods occur in damp areas, too, as this diagram shows.

Heavy rain falls on ground that is already waterlogged
High river due to high rainfall
Rainfall cannot soak into the ground and runs straight off into the river
The river quickly rises and bursts its banks
Extra river pressure feeds into low-lying areas through sink holes

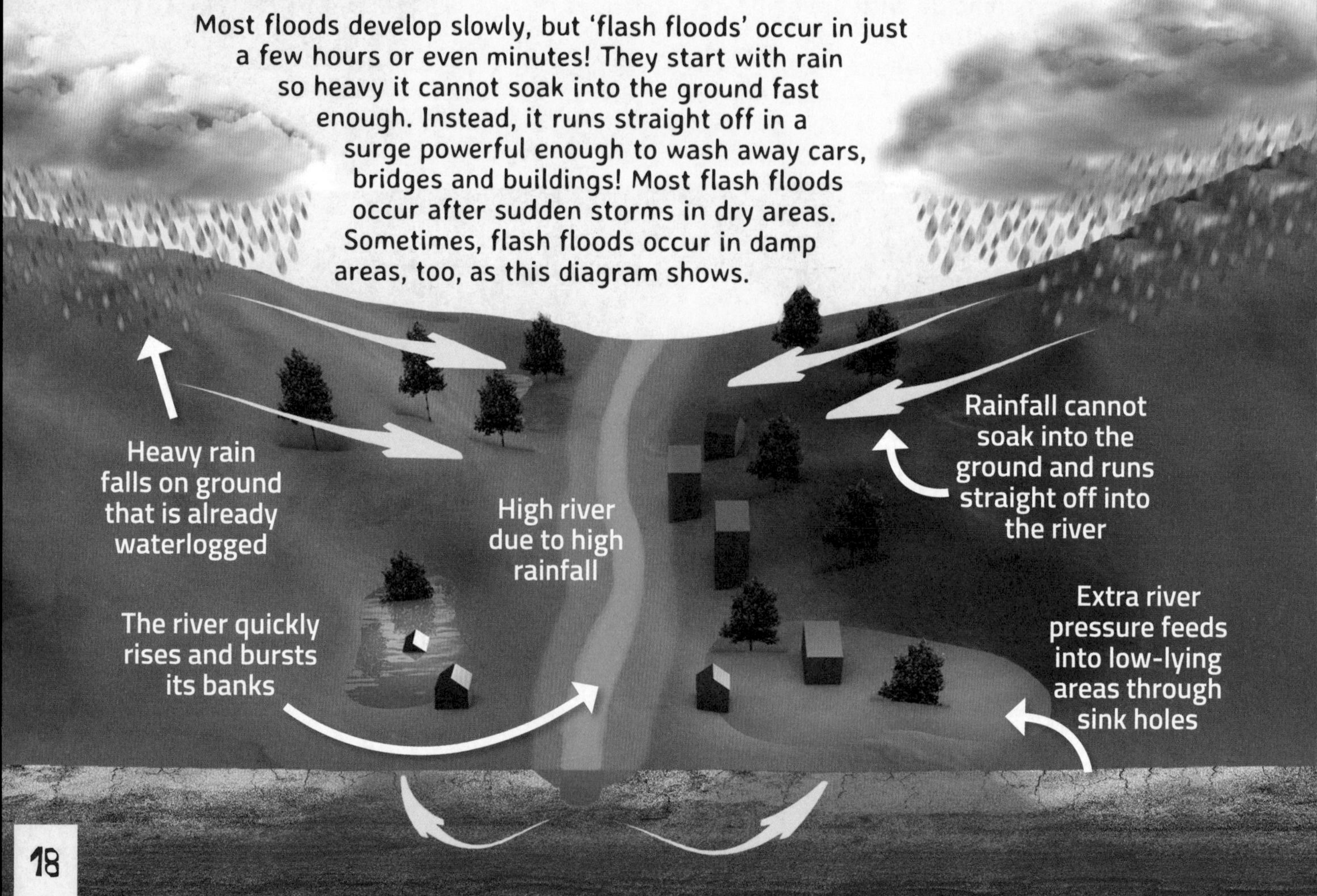

SNOWMELT

In spring, winter snow in the mountains melts and fills local rivers with water. If the temperature rises and melts the snow rapidly, it can unleash dramatic floods. In 1997 in North America, a sudden snowmelt made the Red River of the North rise and swamp vast areas of Minnesota and North Dakota in the USA and Manitoba in Canada.

BLOCKED DRAINS

In most cities, drains are designed to carry rain safely away. But sometimes they get blocked, or rain falls so heavily that drains cannot cope. The result is local floods that can ruin homes and businesses. Occasionally, rain is not the culprit. Water supply pipes can burst and create local floods.

WOOD STRIPPING

Some areas have become more prone to flooding in recent years – and one reason is that trees have been extensively cut down. Trees hold rainwater on the land, and give it time to soak safely into the ground. But if they are felled to make way for farmland or to use the timber, there is nothing to hold the water. Instead it runs straight off, swelling rivers and causing flooding.

EXTREME FLOOD DAMAGE

Floods can have a devastating effect. Terrible floods in China have claimed many millions of lives. But even when people escape alive, homes and businesses can be wrecked and lives torn apart, and the impact can last for years after the flood waters have receded.

UNBELIEVABLE!

The dreadful flood of the Yangtze River in China in August 1931 claimed up to 145, 000 lives. Many people drowned simply because the water rose so high there was no escape. On 19 August, the water level in the town of Hankou in Wuhan rose over 16 m above normal.

EARLY DAMAGE

Floods have many immediate effects. Some people are drowned or stranded and may have to be rescued by boat or helicopter. Buildings and structures such as bridges and roads are damaged or even washed away. Power lines may be damaged, too. Even areas that are not actually flooded suffer.

WATER SHORTAGE

You might think water is the last thing people need in a flood. But flooding disrupts clean water supplies — and the flood waters are usually too contaminated to drink. Flood waters pick up toxic substances as they flow over the land, and become polluted with sewage.

FLOOD DISEASE

Many deaths from floods are caused not by drowning but by diseases that follow. Germs spread in the polluted water and, with little clean drinking water, people succumb to diseases such as cholera. Malaria (being tested for here) can also spread in the stagnant pools left behind.

DOWN BELOW

The lower levels of buildings — basements and cellars — are often the worst-hit by floods. Because they are below ground level, they are hard to drain. Homes with flooded basements are often ruined because it is so difficult to dry them out. The dampness can cause diseases related to breathing.

BROKEN BRIDGES

Bridges span rivers, so are particularly vulnerable to flooding. Some may simply be swamped. Others may be broken as the power of the surging waters washes away the bridge supports. Bridges are vital lines of communication, and the loss of a bridge can trap people, or prevent supplies getting in.

ONE STEP AHEAD OF THE FLOOD

It is not easy to predict when a flood is going to occur. Forecasting the weather is tricky enough. But forecasting when a storm will bring flooding is even harder. All the same, experts are now monitoring rainfall patterns continually and using satellites to help them spot likely flood spots. At the same time, authorities in flood-prone areas are looking at ways of preventing floods.

UNBELIEVABLE!

A 10-year flood is a moderate flood likely to occur every 10 years or so. A huge 100-year flood is likely to occur once every century. But if you're unlucky you could get two 100-year floods in a few years!

STORM WATCH

Hurricanes often bring flooding if they strike the coast in flat, low-lying areas. By analysing past storms, experts can tell how likely flooding is. By tracking an incoming storm by satellite, and using reports from weather stations and other data, weather forecasters can guess how intense it is and where it is likely to hit land. They can then issue flood warnings.

THE WEEK'S RAINFALL AROUND THE WORLD

Little or no rain   Most rain

GLOBAL FLOOD WATCH

The Global Flood Monitoring System, or GFMS, provides a continually updated computer map showing how likely flooding is anywhere in the world. It monitors where rain is falling and how streams are flowing. Satellites also watch where water is building up on land.

FLOODPLAIN

In lowlands, most rivers wind their way through a 'floodplain'. This is a broad, flat area that stretches away on either side of the river. It was built up over thousands of years by silt dropped as the river overflowed its banks again and again, and flooded the landscape. As their name implies, floodplains are very prone to flooding. But a detailed knowledge of the floodplain can tell experts exactly which areas are likely to flood.

FLOOD CONTROL

In places liable to flooding, people often put up barriers to stop the water. Dams may be erected across rivers to hold up the flow at flood times and release the water gradually. Barriers may also be built on the coast to prevent the sea flooding in. Across the River Thames in London, a huge lifting gate called the Thames Barrier (inset) can be lowered quickly to prevent a storm surge out at sea coming up the river and flooding the city.

EXTREME FLOOD STORY

Bangladesh is a low-lying country that is very vulnerable to flooding. Almost three-quarters of the land is less than 8 m above sea level. Floods have struck Bangladesh again and again, but one of the worst events was in 2004. It lasted from July to September and at its worst covered nearly half the country in water.

UNBELIEVABLE!
At the height of the 2004 floods, 40 per cent of Bangladesh's capital Dhaka was under water. The floods killed more than 700 people and made over 30 million homeless.

FLOODING BANGLADESH

There are many reasons Bangladesh is flooded so often. One is its monsoon climate, which means most of its rain is concentrated in one season of the year. Another is the spring melting of heavy snow in the Himalayan mountains. In 2004, the main cause of flooding was a long period of heavy rain that made all three of its big rivers peak at the same time.

Monsoon climate brings heavy rains

Snow melts in the mountains, releasing a lot of water into the river

Much land in the mountains has been cleared of forest, increasing water run-off

Three giant rivers cover 10 per cent of Bangladesh

Soil eroded from areas stripped of forest washes into rivers and chokes them

Urban development has reduced the amount of soft land that can soak up water

Global warming has raised sea levels and increased snowmelt and rainfall

FLOOD RELIEF

This picture shows how used to flooding the people of Bangladesh have become. Women stand patiently in a queue in waist-deep water to collect supplies of food. In 2004, food relief was vital because the floods destroyed millions of tons of rice and drove many people close to starvation.

Although comparatively few people lost their lives in the floods, they caused huge disruption to life. Homes, livestock sheds, vegetable gardens and paddy fields were damaged and destroyed. Many people had to wade through the waters with nothing but their clothes and move into makeshift shelters on higher, drier land or into school buildings, or stay with relatives for months or even years.

STRANDED

The waters rose suddenly, with little warning. Many people were trapped and only escaped by climbing on to the roofs of buildings. They waited for days to be rescued, without food or clean drinking water.

INTENSE FUTURES

In recent years, it seems that wildfires and floods have been becoming more and more frequent and extreme. There have always been serious fires and floods in the past, so this might be coincidence. But many experts are convinced that the way we manage the land is at least partly to blame. So, too, is our changing climate. Are there even more devastating fires and floods to come?

UNBELIEVABLE!

Although global warming is priming forests for more wildfires, most of the recent megafires were started by people. Some were accidents, but many were started deliberately to clear land for farming – then got out of hand.

IF THE ICE SHEETS MELTED...

Some experts believe the Earth could warm up so much that the polar ice sheets could melt entirely. If so, all the water would raise sea levels dramatically. If the Arctic ice sheet melted, sea levels would rise 6 m or so worldwide. If the Antarctic ice sheet melted too, sea levels would rise 60 m! That would drown the USA's east coast, including New York and Florida.

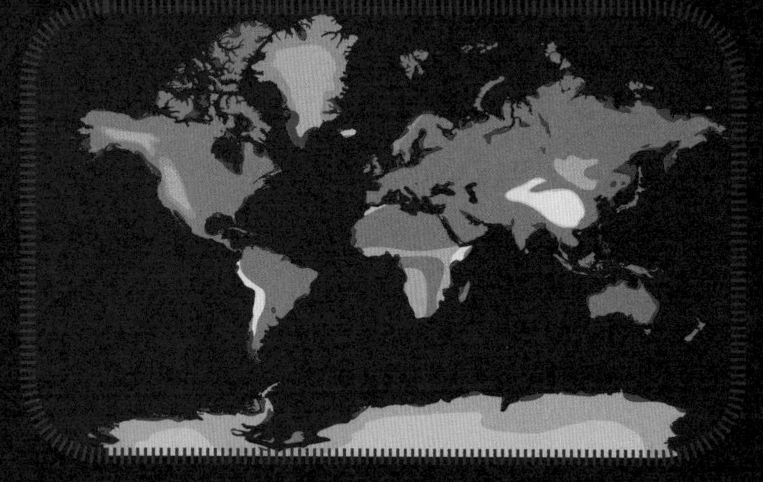

THE AGE OF THE MEGAFIRES

Across the world, devastating wildfires seem to be breaking out more and more often — and they are getting bigger and bigger, too. Experts think the steady warming of the world's climate is to blame for these 'megafires'. The warmer climate makes for longer, drier summers, more fuel for fires as plants grow and dry out, and more lightning as thunderstorms occur more frequently. And, of course, hotter weather primes forests for a conflagration.

LOSING THE MALDIVES

Over the last few decades, as the world has got warmer, the sea has been rising as glaciers and ice sheets melt, and the water in the oceans expands. Experts believe that sea levels are rising by almost 1 cm every year in the Indian Ocean. That doesn't sound much, but it would be enough to drown the beautiful Maldive Islands within the next century, as they are on average just 1 m above sea level.

TIMELINE

1287
St Lucia's flood was a huge storm surge that swamped the Northern Netherlands and Germany and claimed 50,000 to 80,000 lives

1362
St Marcellus's flood was a storm tide that covered vast areas of the Netherlands and created an inland sea, the Zuider Zee

1530
St Felix's flood swamped much of the Netherlands and Belgium and claimed over 100,000 lives

1864
The Great Sheffield Flood claimed the lives of over 240 people in Sheffield, England, in minutes, as the Dale Dyke Dam failed

1871
The Great Peshtigo Fire, the worst in US history, raged through Wisconsin and Michigan

1887
The Yellow River in China flooded, claiming from 900,000 to 2,000,000 lives

1889
In the Johnstown flood, over 2,000 people lost their lives when the South Fork Dam collapsed after heavy rain and flooded Johnstown, Pennsylvania, USA

1910
The Big Burn raged across 12,000 km² in Idaho and Montana, USA, killing 85 people

1911
The Yangtze River in China flooded, taking 100,000 lives

1913
The Great Dayton Flood devastated Dayton, Ohio, USA

1918
The Cloquet Fire in Minnesota, USA, devastated farmland, railway yards and towns, killing 453 people

1931
The Yangtze River in China flooded again, killing up to 145,000 people

1938
The Yellow River in China flooded again in one of the deadliest floods ever, killing 890,000 people

1287

1939
The Black Friday bushfire devastated huge areas of Victoria in Australia

2015
The December floods in the Midwest, USA, caused over $3 billion in damage

1975
The Banqiao Dam in Henan Province, China collapsed after Typhoon Nina, unleashing a wave of water that claimed the lives of over 26,000 people

2009
The Black Saturday bushfire devastated 4,500 km^2 of Victoria in Australia, killing 173 people

2016
Flooding devastated Louisiana, USA, in one of the worst natural disasters of recent years

1971
The Hanoi and Red River Delta flood in Vietnam killed over 100,000 people

2016
The town of Fort McMurray in Alberta, Canada, was lucky to escape with only partial damage after a devastating wildfire

1997
The 1997 Indonesia forest fires were among the worst in modern times

2003
Hundreds of fires broke out across Siberia, releasing huge amounts of CO_2 into the atmosphere

2011
The floods from Hurricane Irene caused huge damage to the USA's Atlantic coast

FIERY AND WET

Amazing facts about fires and floods

HOT LOVE

Fire beetles have their very own heat detectors that use infrared to find burning forest fires. But instead of running away, they head for the fire! Once they're in the thick of it, the beetles mate and lay eggs in the scorched trees. The dead trees are good homes for the beetles, because the sap that usually stops them burrowing has been boiled.

GASSED

One of the big problems for firefighters tackling a big blaze is not the heat but the gases released. The biggest danger is carbon monoxide, which is colourless and odourless and therefore hard to detect. Breathing in carbon monoxide can cause headaches, dizziness, nausea and decreased mental functioning, and may even be fatal.

FIRE STARTERS

Wildfires can start naturally, by lightning or by hot lava. But the US National Park Service fire department believes that up to 90 per cent of wildfires are caused by humans. People may leave campfires unattended, throw away glowing cigarettes or carelessly burn rubbish. Some fires are even started deliberately.

GOOD FLOODS

In many parts of the world, the natural flooding of river plains and deltas each year is a vital part of a farmer's life. The waters bring nutrient-rich silt deposits that create very fertile alluvial soils. In ancient times, many farming communities relied heavily on the annual flooding of floodplain valleys on rivers such as the Nile, the Tigris and Euphrates, and Ganges.

FLASH POWER

The power of moving water in flash floods can be very dangerous. Water moving at just 16 km/h can exert the same pressure as a gust of wind at tornado speed: 434 km/h. The water in flash floods can easily move rocks of 45 kg or more. They can hurl these and other objects at structures with tremendous force – and at people unlucky enough to be caught in the water.

RISING FLOODS

Floods are becoming super-costly in the USA. Between 2011 and 2015, the country was hit by 10 major floods, including those from Hurricane Irene. They caused $34 billion worth of damage.

WORST WILDFIRES

DEADLIEST IN US HISTORY
The Great Peshtigo Fire, Wisconsin/Michigan, USA
October 1871
Cost in lives: at least 1,500
Area: over 15,300 km²

DEADLIEST IN THE LAST 50 YEARS
The forest fires in Indonesia
1997
Cost in lives: 240 people
Area: over 80,000 km²

DEADLIEST THIS CENTURY
The Black Saturday bushfire, Victoria, Australia
7 February 2009
Cost in lives: 173
Area: over 4,500 km²

MOST EXPENSIVE THIS CENTURY
Fort McMurray, Alberta, Canada
May 2016
Financial cost: $2.7 billion

WORST FLOODS

DEADLIEST YELLOW RIVER FLOOD
Central China
September 1887
Cost in lives: 900,000 to 2000,000

SECOND DEADLIEST YELLOW RIVER FLOOD
Central China
1938
Cost in lives: 890,000

DEADLIEST YANGTZE RIVER FLOOD
Central and Eastern China
1931
Cost in lives: 145,000

ONE OF THE WORST DAM BURSTS
Banqiao Dam, Henan, China
August 1975
Cost in lives: over 26,000

INDEX

THE AUTHOR

John Farndon is Royal Literary Fellow at City&Guilds in London, UK, and the author of a huge number of books for adults and children on science, technology and nature, including such international best-sellers as *Do You Think You're Clever?* and *Do Not Open.* He has been shortlisted six times for the Royal Society's Young People's Book Prize for a science book, with titles such as *How the Earth Works, What Happens When...?* and *Project Body* (2016).

Picture Credits (abbreviations: t = top; b = bottom; c = centre; l = left; r = right)
© www.shutterstock.com:
1 c, 9 bl, 11 tr, 14 c, 18 t, 19 tr, 23 c, 23 bl, 29 tl, 30 bl, 32 l.
29 br = © Irene: Cheryl Ann Quigley / Shutterstock, Inc.

NASA:
15 cl

BC, tr = National Geographic Creative / Alamy Stock Photo. BC, cr = Mike Goldwater / Alamy Stock Photo. 2 = AfriPics.com / Alamy Stock Photo. 4 = Sergio Azenha / Alamy Stock Photo. 6, c = Michael Routh / Alamy Stock Photo. 7, c = Mike Goldwater / Alamy Stock Photo. 8 = US Army Photo / Alamy Stock Photo. 10, c = Jennifer Hart / Alamy Stock Photo. 11, cr = National Geographic Creative / Alamy Stock Photo. 12, c = AfriPics.com / Alamy Stock Photo. 13, tr = Tom Bean / Alamy Stock Photo. 13, bc = Sergio Azenha / Alamy Stock Photo. 15, tr = epa european pressphoto agency b.v. / Alamy Stock Photo. 15, cl = NASA. 15, br = epa european pressphoto agency b.v. / Alamy Stock Photo. 16, c = chris Bott / Alamy Stock Photo. 17, cr = MediaWorldImages / Alamy Stock Photo. 19, b = vast natalia / Alamy Stock Photo. 20, c = Global Warming Images / Alamy Stock Photo. 21, t = Ashley Cooper pics / Alamy Stock Photo. 21, c = National Geographic Creative / Alamy Stock Photo. 21, b = dpa picture alliance archive / Alamy Stock Photo. 22, t = Utah Images / NASA / Alamy Stock Photo. 25, tr = REUTERS / Alamy Stock Photo. 25, b = REUTERS / Alamy Stock Photo. 27, cr = imageBROKER / Alamy Stock Photo. 28, br = Chronicle / Alamy Stock Photo. 29, tr = Military first collection / Alamy Stock Photo. 30, tr = Papilio / Alamy Stock Photo. 31 = Eye Ubiquitous / Alamy Stock Photo

极端自然灾害

火灾与洪水

约翰·法恩登（英）著
陈泰之 译

外语教学与研究出版社
FOREIGN LANGUAGE TEACHING AND RESEARCH PRESS
北京 BEIJING

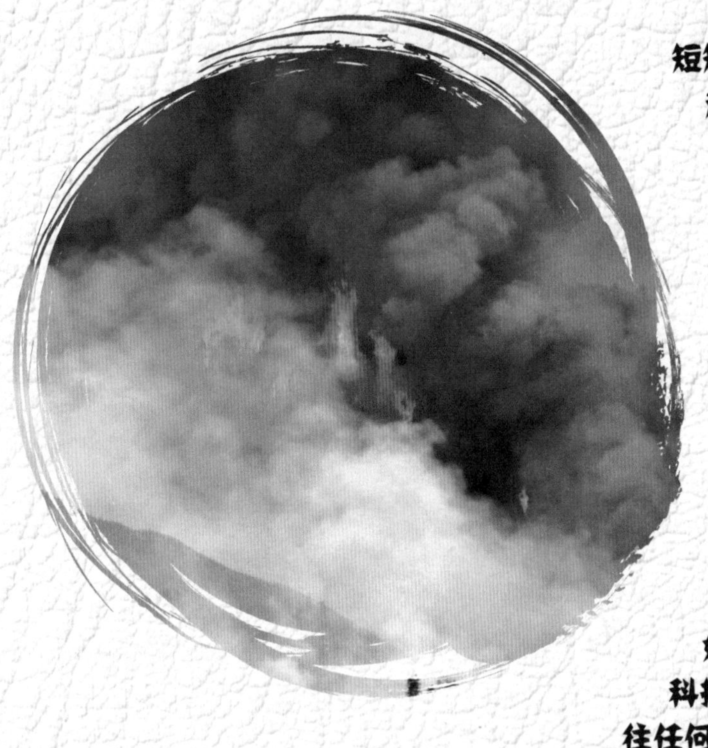

短短数秒，一个火花可以引燃数百处起火点，把大片森林变成人间地狱。然后，控制烈焰的战斗就开始了。史上最大的有些野火让消防队员花了连续数月的时间与之博弈，让我们跟随这些野火，看看它们过后的景象。

水是火的敌人。水对生命至关重要，而由风暴和海平面上升带来的泛滥洪水则是另外一种情况了。看看大洪水如何将陆地变为沼泽，如何席卷房屋和桥梁，如何带来饥饿和疾病。还有，看看科技如何帮助人们预测洪水，这比以往任何时候都更加重要。

目录

野火

野火指火势横扫田野山林，火情失控的大火。它们可以燃烧数日甚至数周，蔓延大片区域，夺去一切生命，威胁附近的居民区。野火可能由人们丢弃的烟头引发，也可能由烧烤野餐的余烬引发，能把整个森林变成一片火海，吞噬挡在它路上的一切东西。

洪水

当大量的水突然淹没旱地时，洪水就形成了。洪水可能在一场强烈的风暴过后迅速发生，然后又眨眼间消退，也可能缓慢地积聚，持续数月。严重的水灾是毁灭性的，能够淹没大片的区域，摧毁人们的家园，经常夺去许多人的生命。

野火是怎样发生的？

当火星引燃干燥的植被时，野火就产生了。野火发生的地方，通常有足够的水分使树木和灌丛生长，又有漫长、温暖而又干燥的夏季。夏季的高温把树木和灌丛烤干，变成大堆木柴和引火物，危险的火花一出现，便会轻而易举地将它们点燃。

你知道吗？

2012年，美国约40,469平方千米的森林发生火灾，面积比美国新泽西州、康涅狄格州与特拉华州的总和还大。现在美国每年森林火灾发生的平均面积是40年前的两倍。

树冠火

野火燃烧的方式分为地下火、地表火和树冠火三种。地下火在土壤中缓慢燃烧，烧掉土壤中的有机物质。地表火沿着地面迅速蔓延，燃烧地面上的落叶与树枝。树冠火是最剧烈的燃烧方式，巨大的火焰席卷了林冠层，烧掉一切可燃物。

火三角

燃烧需要三个条件，也叫做火三角——燃料、氧气和必要的温度。燃料由植被提供，氧气来自空气——燃料燃烧时与空气中的氧气发生反应，必要的温度来自引燃的火星，一旦起火，火焰自身也能提供燃烧需要的温度。三个条件缺一不可，否则火就会熄灭。

火龙卷

野火，尤其是较大的树冠火，在快速吸入空气时会形成风。有时候，尽管只发生在局部区域，这种风却如同飓风一样猛烈。有时，被称为“火龙卷”的龙卷风由野火引发。火龙卷发生时，空气会带着火焰一同螺旋上升。

危险的火星

起火的方式很多。有时可能会自然起火——夏日雷暴时的闪电也许会点燃植被，火山喷发出的灼热灰烬也能引起野火。但许多大火是人为因素引起的——不小心掉落的火柴，或是营火，又或是纵火犯蓄意纵火。

野外大火的危害

野火可能十分危险。对人类来说，危险的不仅是炽热的火焰，还有令人窒息的烟雾。火能够摧毁房屋和烧掉挡在它们路上的几乎所有东西。它们对野生生物，尤其是森林动植物，造成的灾难性影响更甚其他自然灾害。

你知道吗？

一场野火中一般不只有一处火情，因为炽热的物质会随风飘散，引燃多个地方。2009年澳大利亚的“黑色星期六”大火吞噬了大片灌木，这场大火有400处单独的起火点。

烧毁殆尽

野火蔓延迅速，还会快速转向。乘车逃离的人们可能对火情发展的方向做出错误判断，在烟雾中迷失方向，也可能发现逃生路线被烈焰阻断。因此，如果你遭遇到野火险情，在得到官方警报后，一定要尽早撤离。

再生

对于一些植物和动物来说，火情并非灾难，而是机遇。大火清理了朽木，为新生的植物腾出了空间。许多森林在火灾之后迅速恢复了植被的生长。短叶松、五针松和黄桦的球果会受热张开，让种子落入富含草木灰的新鲜营养土壤中。

动物的悲剧

对于许多森林动物来说，野火是一场大灾难。鸟类能够远走高飞，大一些的哺乳动物可以逃离，有些动物会挖洞钻进土中，或者在石头下、在水中躲避野火，但是，动物幼崽和体型小的动物则面临着灭顶之灾，尤其是像松鼠这样的树栖动物。

大火

1910年，大火吞噬了美国爱达荷州、蒙大拿州和华盛顿州约12,141平方千米的土地，夺去了至少85人的生命。城镇被烧毁，数百名士兵前来灭火。自此之后，美国林业局决心扑灭所有林火。然而，现今许多火灾防控专家认为，这一做法不仅给消防队员的生命带来极大危险，而且对保护森林收效甚微。

扑灭野外大火

野火很快就会失去控制，燃烧的区域会十分广阔，要想灭火不仅困难，而且很危险。对待不同的火情要使用不同的灭火方法，但基本的思路是相同的——使大火失去燃料，最终自行燃尽。

目击者

消防员德鲁·米勒通过烧尽燃料的方法灭火，使用“链锯……移除燃料，挖掘防火线使火无法轻易蔓延。然后我们扑灭防火线后易控制的大火，清理火场”。

火情图

计算机和卫星技术使专家们能更密切地关注美国的野火情况。这些技术能够随时实时、交互式地用地图显示何地有野火发生。借助这些手段，美国西北部森林地区的高密度火情也能一目了然。

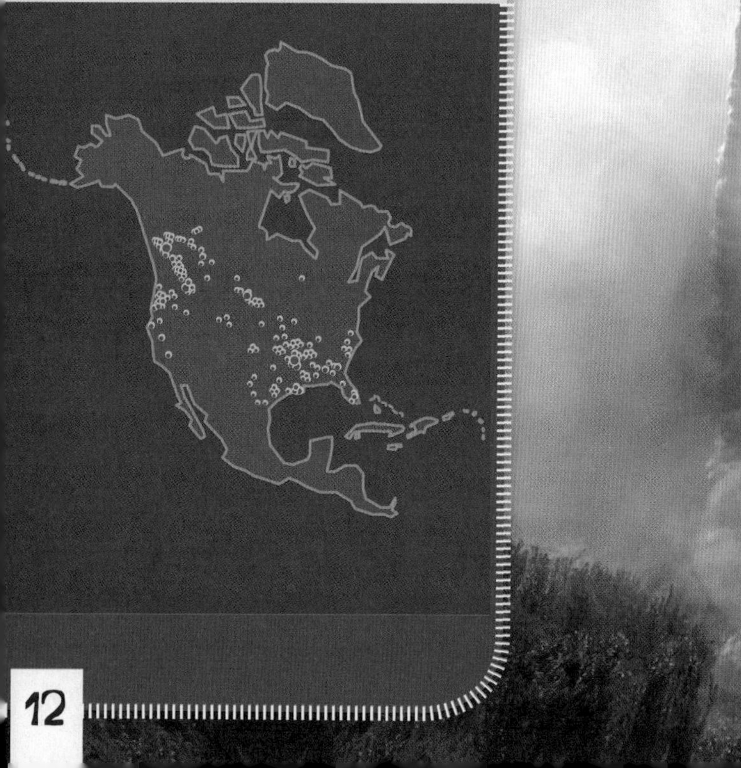

直升机投放

特殊的直升机有时飞临火场上空，投放水、阻燃剂和其他化学物质，比如磷酸铵，以延缓火情。它们一次可以投放7,000升水弹。这样做并非为了灭火，而是要给地面人员争取时间，他们在丛林中设立隔离地带，以最终消灭火患。

观察烟情

通常，在易发野火的森林区域建有瞭望塔，火情观察员可以在瞭望塔上密切关注每一处潜在的火情。他们会追踪并记录下森林里的雷击，持续观察这一区域，搜寻烟雾。

早期警报

过去，突发的火情会让人们大吃一惊，现在，卫星技术的发展使人们能够提前得到火情消息，不必像过去那样惊慌失措。灵敏的卫星能够监测到森林中细微的色调变化，有时这些色调变化表明有烟雾。然后，卫星可以把监测到的信息输入全国野火监控系统，从而向附近居民发出警报。

消防员

地面灭火非常危险，从事这项工作需要经过高强度训练。消防员配备了特殊的防护服装，比如诺梅克斯衬衣和裤子（诺梅克斯材料具有很好的耐高温性，译者注），和在烟雾中呼吸需要的氧气装置。

野外大火的故事

野火灾害可能突如其来。2016年5月1日，加拿大艾伯塔省麦克默里堡突然面临灭顶之灾，被城边森林里燃起的大火彻底摧毁。这场大火只有两人丧生，堪称奇迹，但有2,400座房屋被毁，这场灾难是加拿大历史上损失最惨重的灾难之一，保险赔偿金额约为27.5亿美元。

目击者

消防官员本德费尔德说："我当时在监测一处穿越树林的火情……起初，它的时速是28米。然后它的时速到了32米，再后来达到了300米……怎么控制得住它呢？"

目击者

乔迪·巴茨在外灭火，而他15岁的儿子韦斯利被困在学校，意识到他们或许要失去家园了。以下是他们之间的短信。

巴茨：封闭学校，和大家在一起。要坚强。

韦斯利：高尔夫球场东边的火……是冲着我们家去了吗？

（巴茨知道大火是冲着自己家去了，但他不确定这场火最终会烧到哪里。）

巴茨：火来了。勇敢些。

火焰地狱

在炎热的天气里，浓密、干燥的森林中容易发生火情。但是在麦克默里堡火灾中，风让大火变得特别骇人，也特别难以扑救。大风吹散了炽热的灰烬，在意想不到的地方引燃了新的大火。风抽动着火焰，缓慢前行的一堆火会突然变得烈焰冲天，咆哮着卷过树林。

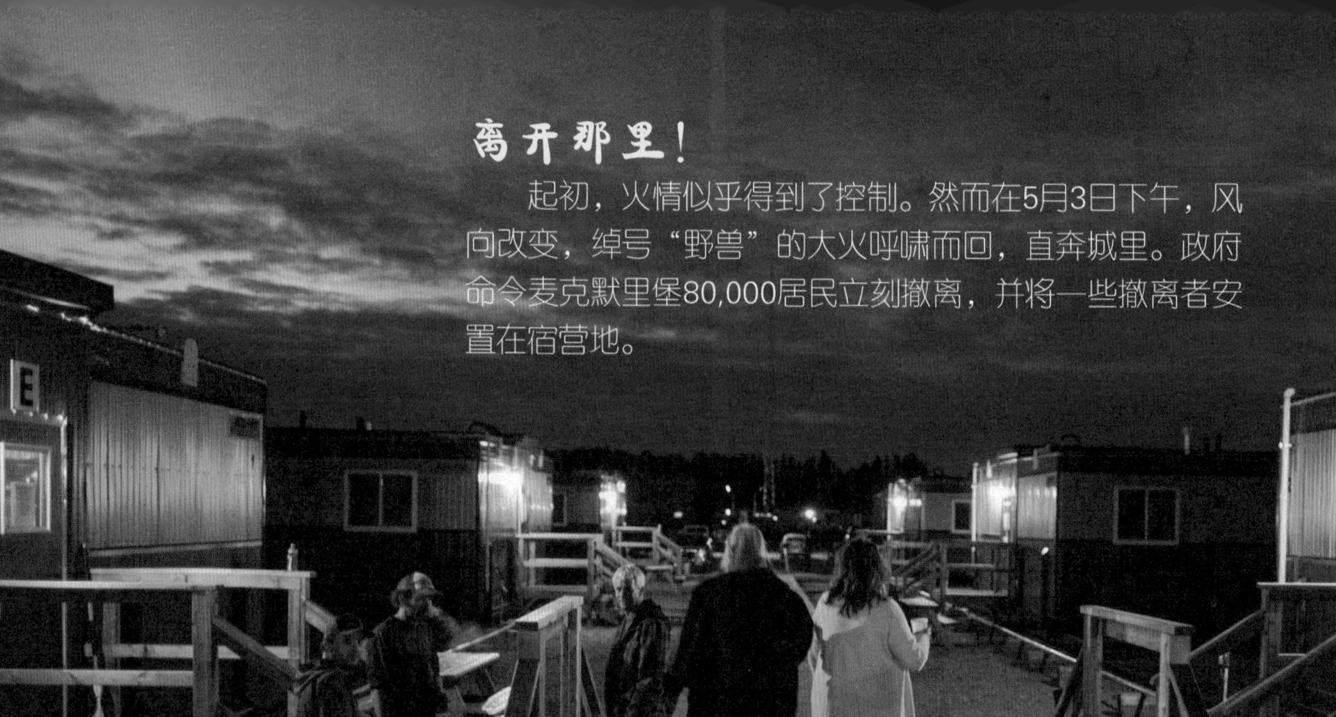

离开那里！

起初，火情似乎得到了控制。然而在5月3日下午，风向改变，绰号“野兽”的大火呼啸而回，直奔城里。政府命令麦克默里堡80,000居民立刻撤离，并将一些撤离者安置在宿营地。

从天上看到的火情！

卫星图像显示，麦克默里堡大火上空烟雾翻滚。到5月中旬，大火蔓延到邻近的萨斯喀彻温省。6月下旬，降雨和冷空气，加上消防人员的出色工作，终于使火情得到了控制。直到2017年8月2日，这场大火才完全熄灭。

路障

大火总是变换着方向，当局不确定该建议人们往哪边撤离。警察设置了路障，阻止人们闯入更危险的区域。路障导致了交通大堵塞。许多滞留在路障处的居民坐在路上，既惊恐又愤怒，而此时他们的家园正在火焰中燃烧。

沿海洪涝

低洼的海岸线很容易遭受来自海上的洪水侵袭。在强风或飓风的作用下，海水会冲向陆地。如果大风同时伴随着强降雨，情况就更糟了。有些破坏力极强的沿海洪涝则是由海啸造成的，海底地震导致巨浪汹涌。

你知道吗？

美国有史以来最高的水位出现在飓风“卡特里娜”发生期间。密西西比州比洛克西的博里瓦奇灯塔处的洪水一度高达10.2米，其中风暴潮6.7米高，其上还有超过3.5米高的浪涛，令人难以置信。

“卡特里娜”洪水

2005年8月，飓风“卡特里娜”袭击新奥尔良期间，风本身并没有造成多大破坏，一天之后到来的洪水才是造成破坏的元凶。飓风带来大量降水，伴随着风暴潮（见右页照片），使纵横全城的水道爆满。最终，防洪的堤坝溃决，洪水灌满了整座城市。

风暴潮

强大的风暴，尤其是飓风，会导致海面突然大幅度升高，这种现象称为风暴潮。风暴潮如同浪高超高的潮汐。风暴中心的低气压使海浪上涌，强风将水吹向岸边，形成波浪。风暴潮能够暂时将海平面抬升数米，淹没沿海地区。

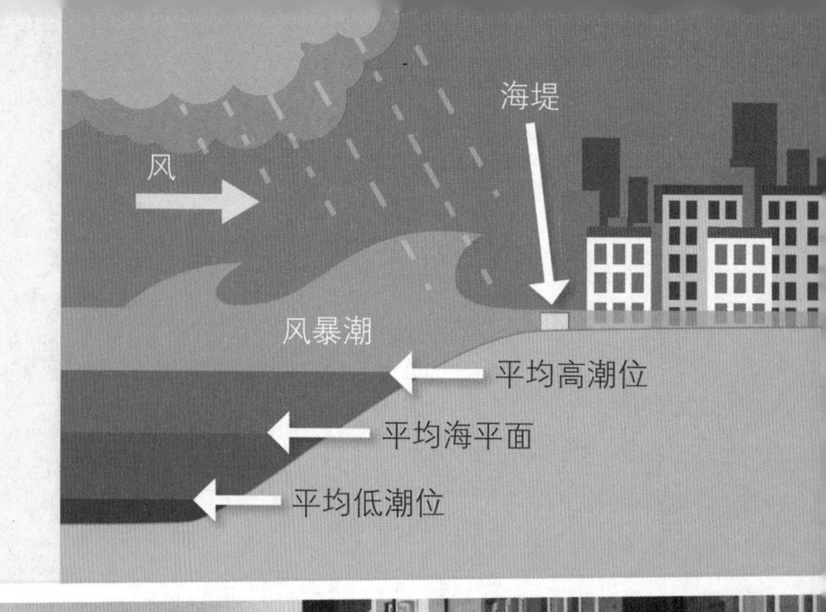

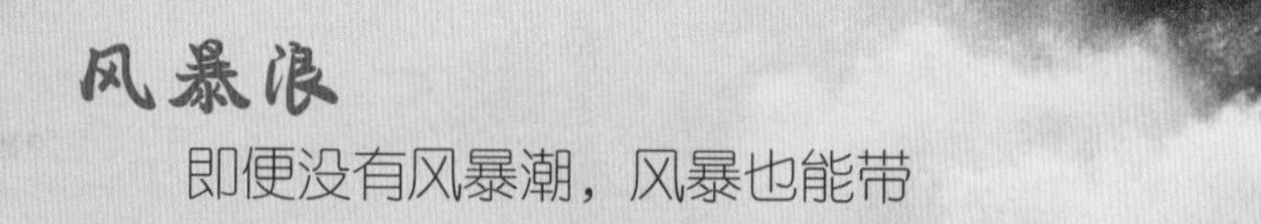

风暴浪

即便没有风暴潮，风暴也能带来大浪。大浪冲击海岸堤防，有时会把堤防冲垮。而当大浪与海水涨潮或风暴潮一同发生时，它们就会形成破坏力巨大的洪水。

高潮

月球绕地球而行，绕一圈需要一个月的时间，这期间海潮的高度也随之变化。每月有两次，月球和太阳排成一条直线。这时海面形成“大潮”，高低潮的潮差最大。这时候，月球和太阳的引力结合起来，极大地拉高和降低海面。当月球和太阳的引力成直角时，海面会形成相对温和一些的“小潮”。

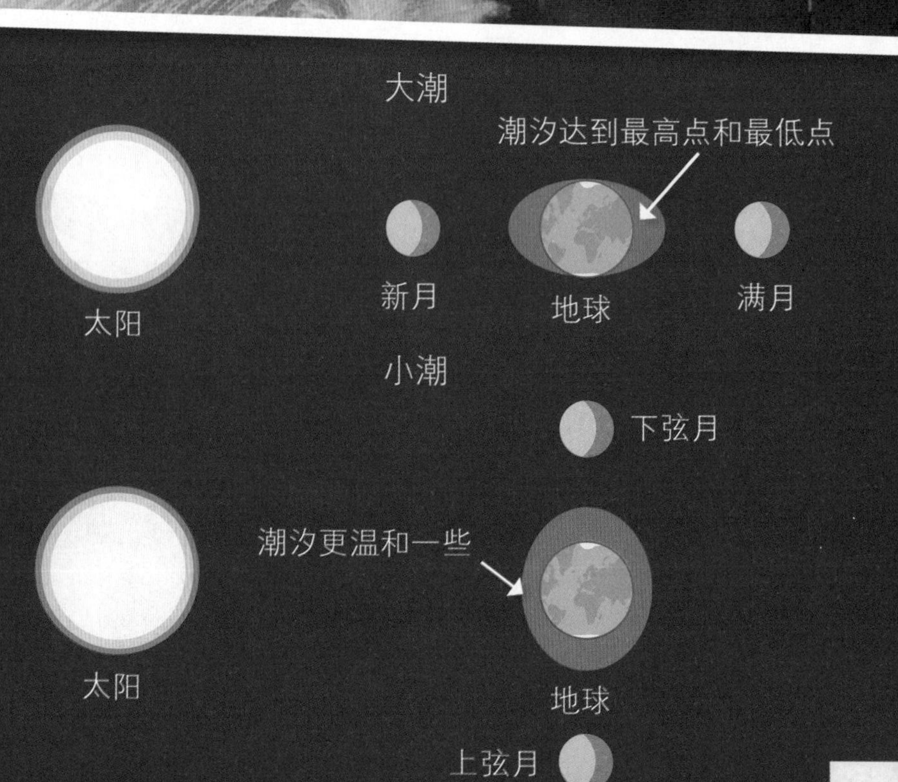

江河洪水

当水量太大，无法通过排水管道和江河安全流走的时候，洪水就发生了。洪水总是出现在大雨或融雪之后。江河湖泊中的水溢流而出，或者雨水太多，无法被土地吸收，于是就在陆地上积聚起来。

你知道吗？

有些最具灾难性的洪水是溃坝造成的。1979年，大雨过后，印度古吉拉特邦“曼朱二号”水坝突然崩坍，大坝后边的湖水倾泻而下。水墙横扫附近的莫尔比，造成25,000人丧生。

暴洪

大多数洪水形成比较缓慢，但“暴洪”发生在几个小时之内，甚至数分钟之间。当雨量太大，雨水无法快速浸入土壤时，暴洪就发生了。它汹涌而来，能冲走车辆、桥梁和建筑！大多数暴洪发生于风暴突袭后的干旱地区。有时，暴洪也发生在潮湿地区，如下图所示。

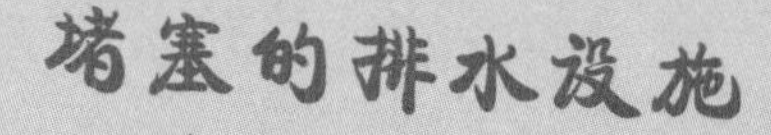

融雪

春天，山区里冬季的积雪融化，雪水注入当地的河流。如果温度快速升高，积雪融化速度太快，就可能发生大洪水。1997年，在北美洲，突然的融雪使北红河河水上涨，淹没了美国的明尼苏达州和北达科他州，以及加拿大的马尼托巴省。

堵塞的排水设施

大多数城市都设计有排水设施，这些设施可以安全地排泄雨水。但有时，排水设施发生堵塞，或者降雨过大，超出排水设施排放能力。这些情况会引发局部洪涝，摧毁住宅和商业设施。有时候，降雨并不是罪魁祸首。供水管线爆裂也能造成局部洪涝。

砍伐树木

近年来，有些地区越来越容易发生洪水，原因之一是过度伐木。在陆地上，树木可以含蓄雨水，使雨水有时间从容地浸入土壤，不会造成危害。如果人们为了开垦农田或者获取木材而砍伐树木，地面上就没有可以蓄水的东西了。这种情况下，水只能流走，充盈江河，引发洪水。

特大洪水的危害

洪水破坏性强。历史上在中国发生的可怕洪水夺去了数以百万的生命。即使人们得以逃生，家园和生计也被毁坏，生活被撕裂，洪水退去多年之后其影响仍然存在。

你知道吗？

1931年8月，骇人的长江洪水夺去了145,000人的生命。许多人直接溺水而亡，因为水位涨得太高，他们根本没有逃生之处。8月19日，武汉市的汉口水面高度较正常水位高出16米。

早期损失

洪水带来的影响，不少会即时显现。有人被淹，有人被困，必须靠船只和直升机才能获救。楼房和桥梁、道路等建筑物受损，甚至被冲走。输电线路也会受损，因此，并未实际遭遇洪水的地区也可能受到影响。

缺水

你也许觉得，洪水中人们最不缺的东西就是水。然而，洪水破坏了清洁水的供应，而洪水本身通常污染严重，无法饮用。洪水从地上流过时会吸收有毒物质，并被污物污染。

洪水过后的疾病

许多人因洪水死亡，但他们并不是被水淹死的，而是死于洪水过后的疾病。细菌在受污染的水中传播，由于缺乏清洁的饮用水，人们会死于霍乱之类的疾病。洪水过后会留下死水池，疟疾（图中正在进行疟疾检验）会在其中传播。

低洼处

建筑物的较低层——地下室和地窖——通常是被洪水损毁最严重的地方，因为它们低于地面，很难排水。如果地下室被水淹，那么住宅常常随之被毁，因为人们很难将水排干。潮湿能够引发与呼吸系统有关的疾病。

断桥

桥梁横跨河流，因此特别容易受到洪水的影响。有些桥梁只是被淹没，其他一些桥梁则可能被冲断，因为它们的桥梁支座被汹涌的洪水冲垮了。桥梁是交通生命线，失去桥梁，人们会被困住，物资供给也会受阻。

洪水发生前的预警

预测何时发生洪水不是一件易事。预报天气已经够复杂了，预报风暴何时造成洪水更是难上加难。尽管如此，专家们一直在监测降雨模式，并且利用卫星等辅助手段发现可能发生洪水的地点。同时，易发洪水地区的行政当局也在寻找防止洪灾的方法。

你知道吗？

十年一遇的洪水是中等程度的洪水，可能十年左右发生一次。百年一遇的大洪水可能每个世纪发生一次。但如果不走运的话，你可能在几年里遇到两个“百年一遇”的洪水！

风暴监测

如果飓风袭击了平坦、低洼的海岸地带，通常会造成洪涝灾害。通过分析以往的风暴，专家能够判断发生洪水的概率。借助卫星追踪来袭的风暴，再参考气象台的报告和其他数据，天气预报员能大致预估它的强烈程度，推断受灾的具体位置，然后他们就能发布洪水警报。

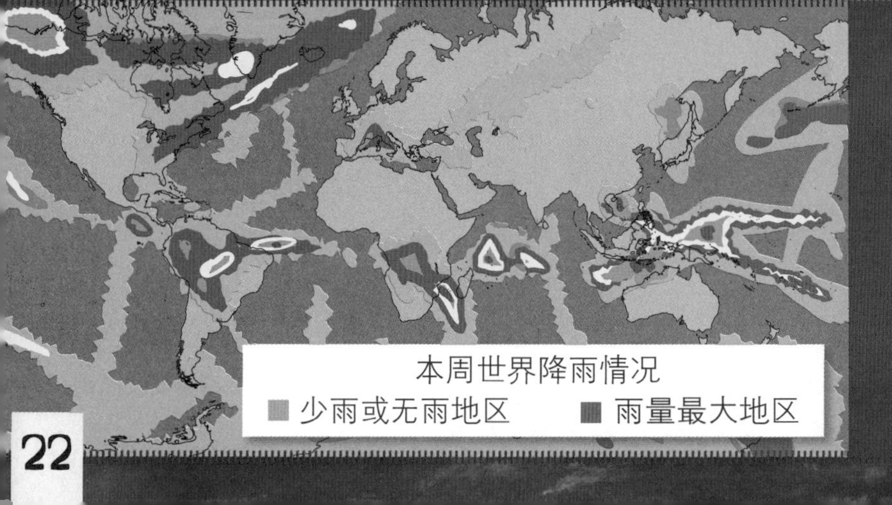

全球洪水监测

全球洪水监测系统（GFMS）提供不断更新的计算机电子地图，显示世界上各个地区发生洪水的可能性。该系统监测着世界各地的降雨和水流情况。卫星还注意观察哪些区域有水流积聚的现象。

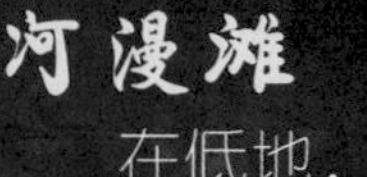

河漫滩

在低地，大多数河流蜿蜒穿过“河漫滩”。河漫滩是河床两侧的平坦地带。它是在数千年的时间里，由于河水不断溢出河岸、淹没陆地，泥沙淤积而形成的。顾名思义，河漫滩是洪水易发之处。而对河漫滩的详细了解可以让专家知道，到底哪些地区可能会洪水泛滥。

模拟洪水深度（米）

<0.5

0.5 – 1.0

1.0 – 2.0

2.0 – 4.0

河谷全长

防洪

在易于发生洪灾的地区，人们经常建起屏障以阻止洪水。水坝跨河而建，在洪水时期可以蓄洪，再逐步将水排出。防洪堤坝也可能筑于海岸上，抵御海潮。在伦敦的泰晤士河上，称作泰晤士河防潮闸的巨大抬升式门型建筑（见上方小图）可以迅速放下，以防止海面上的风暴潮溯河而上，淹没城市。

特大洪水的故事

孟加拉是个低洼的国家，很容易遭受洪涝灾害。该国将近四分之三的土地海拔高度不超过8米。洪水一再冲击孟加拉，最严重的一次发生在2004年。水灾从7月持续到9月，最严重时全国近一半地区淹没在水中。

你知道吗？

在2004年洪水最凶猛时，孟加拉首都达卡40%的地区被大水淹没。洪水夺去700多人的生命，导致3,000万人无家可归。

遭受洪水灾害的孟加拉

孟加拉经常遭受洪水袭击，原因很多。一个原因是它的季风气候，这意味着该国的降水集中于一年中的某个季节。另一个原因是喜马拉雅山春季的大量融雪。2004年洪涝灾害的主要原因是，长时间的大雨使孟加拉境内三条大河同时达到了水量顶峰。

季风气候带来强降雨

山上积雪融化，大量雪水汇入河流

山区大片土地上的森林被砍伐，造成河水溢流现象加剧

三条大河的流域面积占孟加拉国土面积的十分之一

有些地区的森林遭受砍伐，土壤被冲刷进河里，阻塞了河道

软土能够吸收水分，但城市发展减少了软土面积

全球变暖使海平面升高，增加了融雪和降水量

洪水中的救济

右侧照片显示，孟加拉人已经十分习惯于洪水了。妇女们耐心地站在齐腰深的水中，排队领取食物。2004年，食品救济至关重要，因为洪水毁掉了数百万吨稻米，使许多人处于饥饿边缘。

相对来说，洪水本身没有造成多少人死亡，但却给人们的生活造成巨大困扰。人们的家园、畜舍、菜园和水稻田都遭到破坏。许多人只能带着随身衣物，蹚过洪水前往位置更高、较为干燥处的临时避难所，或者躲进学校，有些则和亲戚一起待上几个月，甚至几年。

受困

水位毫无明显征兆地突然升高，许多人被困，只能爬上屋顶躲避。他们要等待多日才能获救，这些天里没有食物，也没有清洁的饮用水。

严峻的未来

近年来，野火和洪水似乎越发频繁，也越发极端。其实，过去也常发生严重的火灾和水灾，所以目前的状况可能只是巧合。但许多专家坚信，我们使用土地的方式对灾情的发生有一定的责任，气候变化也是部分原因。未来还会有更可怕的火灾和洪水发生吗？

你知道吗？

虽然全球变暖让森林更易发生野火，但是最近发生的多数大火是人为造成的。这些大火中，有些是意外事故，还有很多则是为了开辟耕地而人为引燃，并发展成了失控的大火。

如果冰盖融化……

有些专家认为，地球变暖的程度可能足以使极地冰盖完全融化。倘若如此，融化的水会使海平面显著升高。如果北极冰盖融化掉，全世界的海平面会升高约6米。如果南极冰盖也融化掉，海平面会升高60米！那么美国东岸就会被淹没，包括纽约和佛罗里达等地。

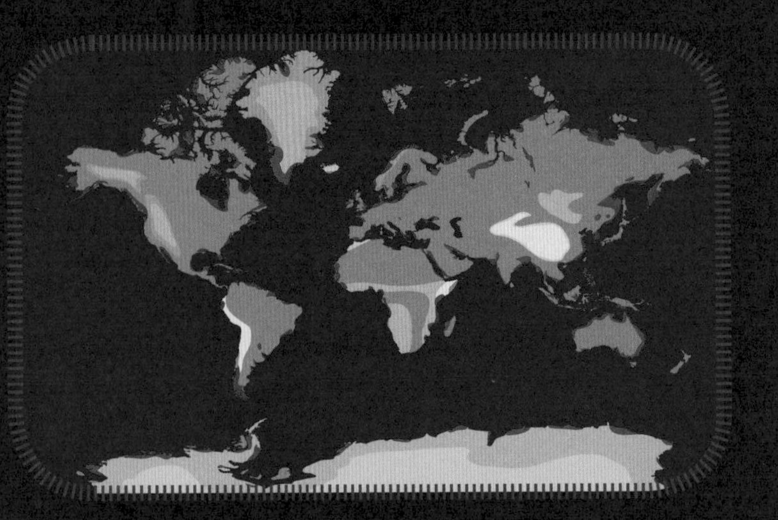

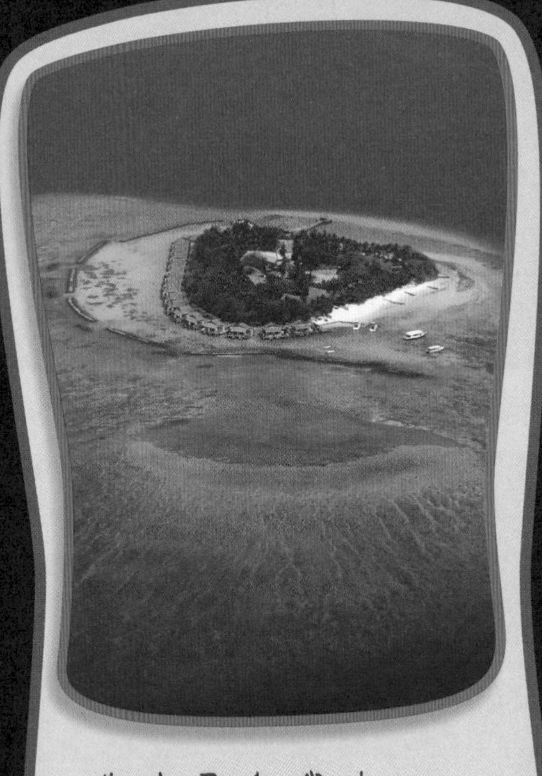

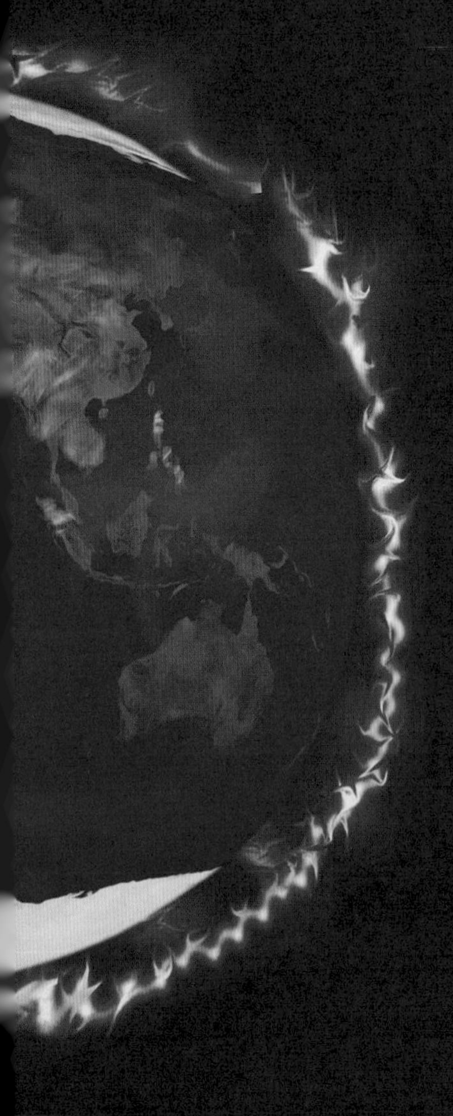

大火的时代

世界上灾难性的野火发生得越来越频繁，规模也越来越大。专家认为，这些“大火”归咎于世界范围内持续的气候变暖。气候变热，使得夏季更长、更干燥；植物因干旱天气而变枯，为野火提供了更多的燃料；雷暴发生的频率增加，带来了更多雷电。当然，炎热的天气使森林成为火灾的多发区。

失去马尔代夫

过去几十年间，全球变暖，冰川和冰盖融化，海洋的水域扩展，海面也在升高。专家相信，印度洋的海平面每年升高近1厘米。这听上去似乎不算多，但按照这个速度下去，美丽的马尔代夫群岛在下个世纪就会被完全淹没，因为这个群岛的平均海拔不过1米。

大事件时间轴

1287年
由一场巨大的风暴潮造成的圣卢西亚洪水淹没了荷兰北部和德国，夺去了50,000到80,000人的生命

1362年
圣马塞勒斯洪水是一个风暴潮，淹没了荷兰许多地区，形成了一个内陆海——须得海

1530年
圣费利克斯洪水席卷荷兰大部分地区和比利时，夺去超过100,000人的生命

1864年
戴尔戴克大坝垮塌，谢菲尔德大洪水在数分钟内夺去英格兰谢菲尔德市240多人的生命

1871年
美国历史上危害最大的佩什蒂戈大火横扫威斯康星州和密歇根州

1887年
中国黄河发生了大洪水，夺去900,000至2,000,000人的生命

1889年
大雨过后，南福克大坝垮塌，洪水淹没美国宾夕法尼亚州约翰斯敦，2,000多人失去生命

1287

1910年
“大火”横扫美国爱达荷、蒙大拿两州12,000平方千米土地，导致85人死亡

1911年
中国长江水患夺去100,000人的生命

1913年
代顿大洪水侵袭美国俄亥俄州代顿市

1918年
美国明尼苏达州克洛凯大火焚毁了农田、铁路货场和城镇，夺去453人的生命

1931年
中国长江再次泛滥，死亡人数多达145,000人

1938年
中国黄河发生中国历史上最致命洪水之一，导致890,000人死亡

1939年

“黑色星期五”森林大火将澳大利亚维多利亚州大片地区摧毁

2015年

美国中西部发生的“12月”大洪水造成逾30亿美元的经济损失

1975年

台风“尼娜”过后，中国河南省板桥水库溃决，洪水夺去逾26,000人的生命

2009年

“黑色星期六”大火肆虐澳大利亚维多利亚州4,500平方千米土地，大火造成173人死亡

2016年

美国路易斯安那州发生的洪涝灾害为近年最大的自然灾害之一

1971年

在越南河内和红河三角洲发生的洪水造成逾100,000人死亡

2016年

一场破坏性极大的野火过后，加拿大艾伯塔省麦克默里堡很幸运地只有部分地区在大火中受损

1997年

1997年印度尼西亚森林大火是现代史上影响最恶劣的火灾之一

2003年

西伯利亚地区出现了数百个着火点，释放的大量二氧化碳气体进入大气层

2011年

飓风“艾琳”带来的洪水给美国大西洋沿岸地区造成巨大损失

嗜热

火甲虫有其独特的热量探测器，能够使用红外线寻找森林中的起火点。遇到火时，火甲虫们不会逃之夭夭，反而会奔向起火点！一旦到达起火地点，它们便会交配，并在烧焦的树木上产卵。对甲虫来说，死树是安家的好地方，因为通常阻碍它们挖洞的树液已经烧干了。

毒气中毒

消防队员在应对大火时，面临的一个重大问题不是高温，而是火灾中释放出的气体。最危险的气体是一氧化碳，因为一氧化碳无色无味，很难被人发觉。一氧化碳能让人头疼、头晕、恶心、神志不清，甚至可能置人于死地。

起火原因

野火可能是自然原因引起的，由闪电或热熔岩引发。但美国国家公园管理局消防部门认为，高达90%的野火是由人类造成的，例如人们离开宿营地时没有熄灭营火，人们随手扔掉未熄灭的香烟，或者粗心大意地用火焚烧垃圾。有时，甚至有人故意纵火。

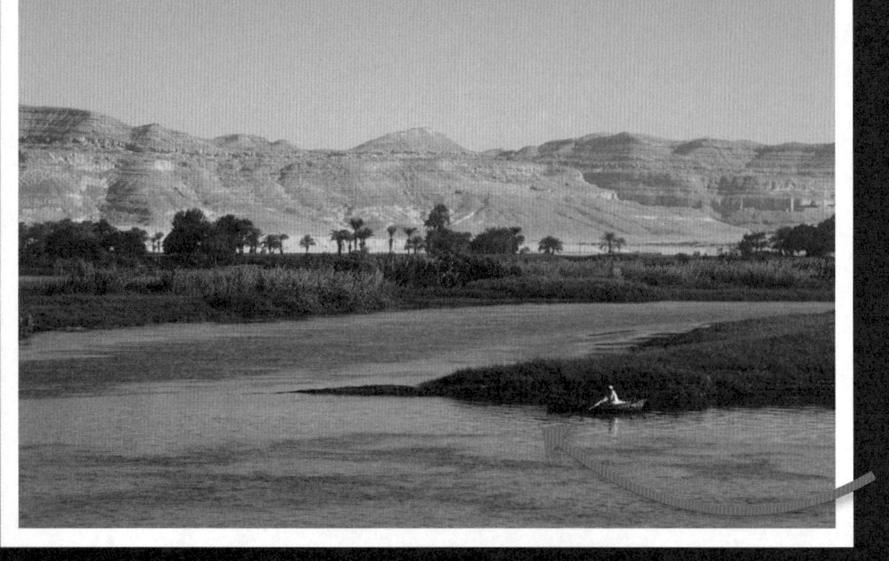

有益的洪水

在世界上的许多地方，江河每年在冲积平原和三角洲地区的自然泛滥对农民的生活十分重要。河水带来营养丰富的淤泥沉积物，形成十分肥沃的冲积土。在古代，很多农耕群落非常依赖于每年泛滥的河漫滩谷地，产生这些谷地的河流有尼罗河（世界第一长河，流经非洲东部和北部，译者注）、底格里斯河与幼发拉底河（这两条河流起源于土耳其，流经叙利亚和伊拉克，注入波斯湾，译者注），以及恒河（流经印度的一条大河，译者注）。

暴洪的威力

暴洪中的水流是十分危险的。时速16千米的水流产生的压力，相当于时速434千米的狂风行进时产生的压力，这个风速已经达到龙卷风的速度了。暴洪发生时，水能轻易地移走45千克或更重的石头，能以惊人的力量把石头和其他物体抛向建筑物——或砸到不幸被困在水中的人身上。

不断增长的洪涝损失

在美国，洪涝灾害造成的损失十分巨大。2011至2015年间，全国遭遇10次大洪水，包括飓风“艾琳”引起的洪水，总计损失约340亿美元。

野火之最

美国历史上死亡人数最多
美国威斯康星州和密歇根州的佩什蒂戈大火
1871年10月
死亡人数：至少1,500人
受灾面积：15,300平方千米

近50年内死亡人数最多
印度尼西亚森林大火
1997年
死亡人数：240人
受灾面积：超过80,000平方千米

本世纪死亡人数最多
“黑色星期六”大火
澳大利亚维多利亚州
2009年2月7日
死亡人数：173人
受灾面积：超过4,500平方千米

本世纪损失最大
加拿大艾伯塔省麦克默里堡大火
2016年5月
经济损失：27亿美元

洪水之最

死亡人数最多的黄河洪水
中国中部
1887年9月
死亡人数：900,000至2,000,000人

死亡人数第二的黄河洪水
中国中部
1938年
死亡人数：890,000人

死亡人数最多的长江洪水
中国中部和东部
1931年
死亡人数：145,000人

最严重的溃坝事件之一
中国河南板桥水库
1975年8月
死亡人数：超过26,000人

索引

作者简介

约翰·法恩登，英国伦敦城市行业协会的皇家文学基金会研究员。他为成人与儿童写作了大量图书，涉及科学、技术、自然等领域，包括畅销世界的《你觉得自己聪明吗？》和《放我出去》。他六次入围英国皇家学会青少年科学图书奖，入围书目有《地球原理》《何因何果》和《人体项目》（2016）等。

WHEN DISASTER STRIKES
EXTREME VOLCANOES

Written by John Farndon

外语教学与研究出版社
FOREIGN LANGUAGE TEACHING AND RESEARCH PRESS
北京 BEIJING

Our restless, fiery Earth can erupt suddenly and violently in very many places at any moment. In fact, there are at least 20 eruptions happening right now! They can bury entire cities in ash, cause scorching avalanches, poison the air, trigger tsunamis and drastically alter our climate.

Witness the worst that volcanoes have done and may do, see how they occur and where the hot spots are, why people live near them, and what those clever volcanologists are doing to predict future explosions...

CONTENTS

VOLCANOES

A volcano is a place on the Earth's surface where molten rock from its hot interior bursts through its solid crust. Sometimes, it is just a crack where the molten rock, known as magma, oozes through as lava (which is the name for magma on the surface). Sometimes, though, it is a cone-shaped mountain built up by successive eruptions. Mountain volcanoes can erupt in sudden and dramatic explosions, unleashing huge amounts of scorching hot material in one blast.

7

HOW DOES A VOLCANO HAPPEN?

Beneath most mountain volcanoes there is a reservoir of hot magma called the magma chamber. Over time, pressure builds up in the chamber as magma builds up from beneath. Eventually, the pressure is enough to drive the magma out onto the surface through the volcano's narrow chimney or 'vent'.

EYEWITNESS

Plinian eruptions are named after the terrible eruption of Vesuvius in Italy in 79 AD, which killed Roman scholar Pliny the Elder and was witnessed by his nephew from a boat at sea.

Lightning strikes caused by a build-up of static electricity are common in the most violent eruptions

The eruption blasts out fragments of the clog in the vent in a cloud of ash and debris

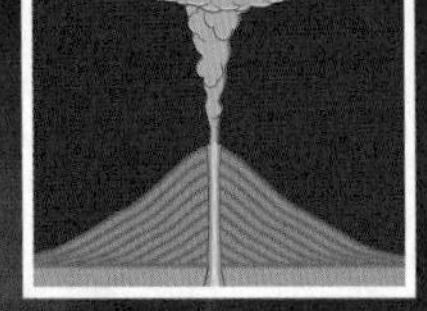

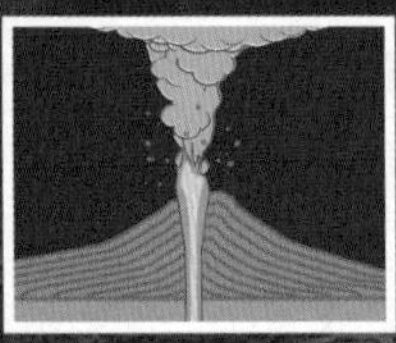

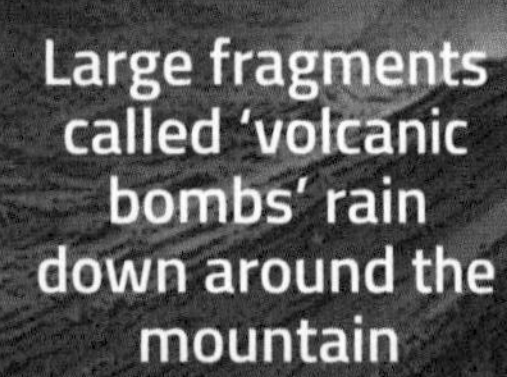

Large fragments called 'volcanic bombs' rain down around the mountain

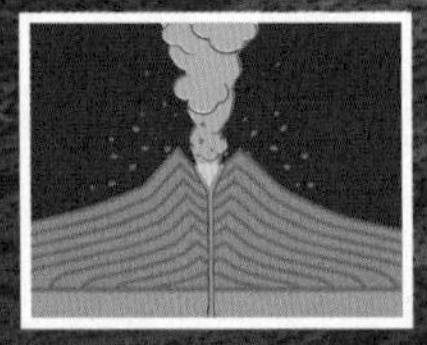

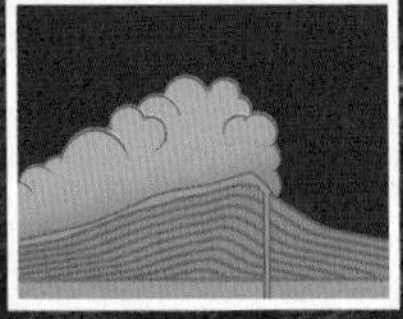

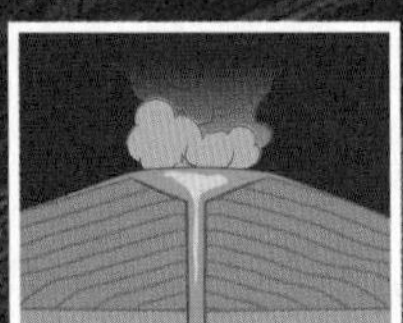

Sometimes, magma may burst through the side of the volcano to form smaller 'secondary cones'

Main vent

Once the ash and debris is blasted out of the way, molten magma floods out and runs down the mountain as streams of lava

THERE SHE BLOWS!

In an explosive volcano, the vent is clogged up with solidified material from previous eruptions. A new eruption begins when rising magma pushes its way through the clog. As the magma bursts through, the pressure in it drops. Bubbles of carbon dioxide gas and steam suddenly form and drive the magma whooshing out like a fizzy drink from a shaken bottle.

The magma chamber, where magma builds up beneath the volcano before an eruption

KINDS OF ERUPTION

These are just some of the different kinds of volcanic eruption:

Plinian eruptions are the most violent, rocketing gas and volcanic ash high in the air, where they spread out far and wide in a flat-topped cloud.

Vulcanian eruptions produce less violent explosions of gas that are laden with volcanic ash and form dark, churning clouds. These billow quickly in convoluted shapes.

Strombolian eruptions occur in quick bursts, in which ash mixes with glowing globs of hot lava.

Pelean eruptions throw out high-speed avalanches of scorching hot ash and debris called pyroclastic flows.

Hawaiian eruptions are less violent but more continuous, emitting floods of runny lava that flow far out, cool and solidify to form shallow 'shield' volcanoes (see p. 10).

FAST AND FURIOUS

Super-explosive 'plinian' eruptions send out giant blast after giant blast of gas and volcanic ash, along with hot fragments of bubbly solidified lava called pumice. They rarely last more than a few weeks and sometimes less than a day.

BIG BANG

The explosive eruption of Mount Pinatubo, in the Philippines on 15 June 1991 was one of the biggest eruptions in recent history. It sent out high-speed avalanches of hot ash and gas (pyroclastic flows), giant mudflows (lahars) and a cloud of volcanic ash that was 400 km in diameter.

EXTREME VOLCANO DAMAGE

Volcanic eruptions can do terrible damage in many ways. They can send out suffocating clouds of gas and ash. They can bomb the surrounding area with red-hot fragments. They can belch unstoppable streams of red-hot lava, and scorching avalanches that destroy everything in their path. They can even set off earthquakes and landslides.

EYEWITNESS

Joseph Nkwain was at Cameroon's Lake Nyos disaster in 1986 (below right). "I could not speak. I became unconscious. I could not open my mouth because then I smelled something terrible..."

KINDS OF VOLCANO

Caldera complexes have a 'caldera' (crater), from the mountain sinking into the magma chamber, like those in Yellowstone Park.

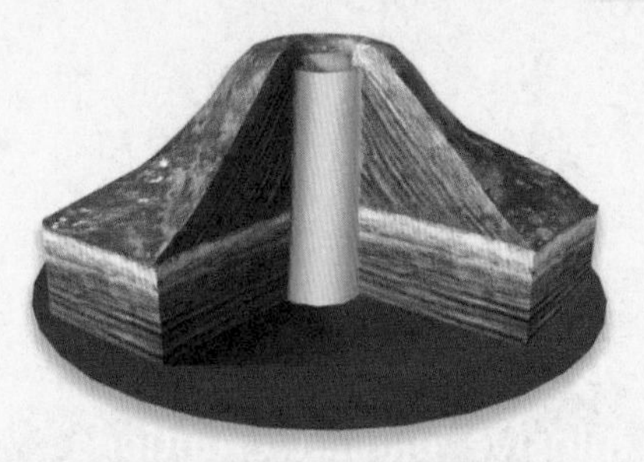

Dome volcanoes form from sticky lava glugging out into a dome-shaped mound. France's Puy-de-Dôme is an extinct dome.

Fissure volcanoes are long cracks in the ground where lava oozes out continually with little explosive activity, as in Iceland.

Shield volcanoes form from runny lava flowing out far and wide. Hawaii's Mauna Kea is a shield volcano.

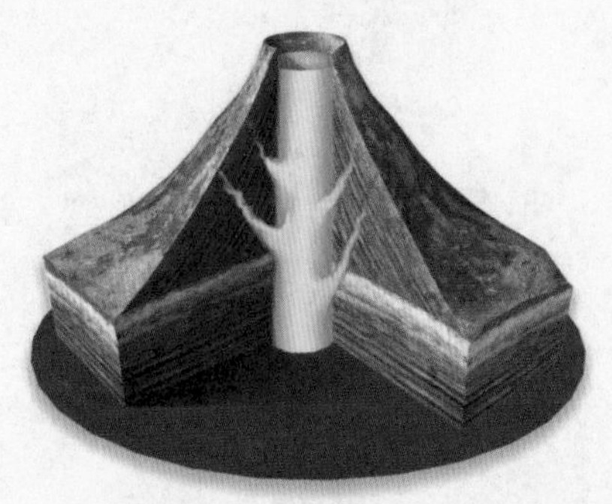

Stratovolcanoes or 'composites', such as Mount Etna in Sicily, Italy, are made of layers of lava and ash from successive eruptions.

Ash and cinder cone volcanoes are cone-shaped mountains formed from ash and cinder. Krakatoa in Indonesia is a cinder cone.

MUDSLIDE

Some eruptions can set off devastating flows of scalding rock, mud and water called lahars. They typically start when the heat of an eruption melts snow or frozen ground. The mud hurtles downhill as fast as a car. When a lahar hit the city of Armero after the Nevado del Ruiz eruption in Colombia in 1985, 25,000 people were killed.

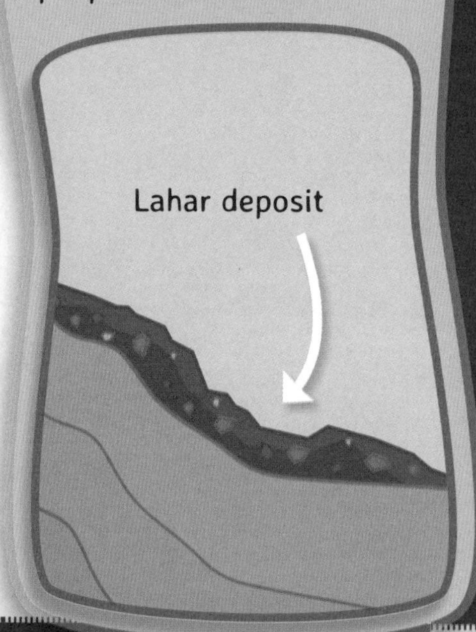

A VOLCANO'S ARMOURY

Volcanoes can throw up many kinds of danger!

- Lava flows – streams of red-hot, molten rock – are terrifying and unstoppable, but usually move slowly enough for people to run away. But they destroy buildings.
- Scorching, choking clouds of ash and cinder falling far from the volcano are the big danger and there is little chance of escape. The weight of ash makes roofs collapse.
- Fine dust formed from volcanic debris can float gradually down and cause slow suffocation.
- Pyroclastic flows are superhot, high-speed avalanches of ash, cinders and gas.

GASSED!

In one of the eeriest volcanic disasters, more than 1,700 people and many animals were killed invisibly within minutes in Cameroon on 21 August 1986. The cause was a cloud of suffocating carbon dioxide gas released from nearby Lake Nyos. The lake was pumped full of the dangerous gas by mild volcanic eruptions from below the lake.

THE MOST DANGEROUS VOLCANOES

Getting caught near an erupting volcano is very dangerous. Yet many people live near volcanoes that have not erupted for so long they seem safe to live near. But are they really 'extinct' (stopped completely) or 'dormant' (sleeping) – or just biding their time until the next devastating eruption?

SHOCK ERUPTION

In 1980, most people thought Mount St. Helens in Washington was dead, since there had been no activity for over 100 years. But on 20 March that year, it suddenly exploded with the deadliest and costliest eruption in US history, killing 57 people.

ACTIVE VOLCANOES

There are no firm rules about what makes a volcano active, dormant or extinct. But when you can see smoke or lava pouring out, as in Kilauea in the USA, Sakurajima in Japan, or Cotopaxi (above) in Ecuador, you know it's active! Volcano scientists usually say that if it's erupted in the last 10,000 years, it's active.

DORMANT VOLCANOES

A dormant volcano is one that hasn't erupted for 10,000 years, yet still might erupt one day. Mount Kilimanjaro in Africa hasn't erupted for 360,000 years – but it's not entirely dead yet.

EXTINCT VOLCANOES

Extinct volcanoes have not erupted for tens or hundreds of thousands or even millions of years. They seem unlikely to erupt ever again because the magma beneath has gone cold and solid. Edinburgh Castle in Scotland is famously built on an extinct volcano.

ACTIVE OR SLEEPING?

Mount Fuji in Japan is one of the world's most famous volcanoes. It hasn't erupted for over 300 years. So is it sleeping, or is it preparing for a surprise? Small holes, called fumaroles, on the mountain release puffs of smoke even now. No one is sure if they are a danger sign or not.

SLEEPING GIANT

Mount Rainier in Washington looks calm and beautiful with its snowcapped peak. It hasn't erupted since 1894, so might seem to be sleeping. But a lot of people live nearby, and all that ice could melt in a terrible avalanche if it suddenly erupted.

EXTREME WEATHER EFFECTS

Volcanoes don't only make their impact felt on the ground and the landscape. They can change the atmosphere, too. When an explosive volcano erupts, it spews out so much ash and polluting gas into the atmosphere that this can actually change the weather not just locally, but right round the globe.

UNBELIEVABLE!

When Huaynaputina in Peru erupted on 19 February 1600, it claimed two million victims far away in Russia, where the ash thrown out led to a summer so cold it brought a terrible famine.

VOLCANIC STORM

As if you weren't frightened enough by all the debris and lava chucked out, a volcanic eruption often creates a violent storm, too! All the ash particles blasted high in the air attract water droplets. So the eruption becomes just like a super-intense thundercloud, with lots of rain, thunder and lightning flashes.

NASTY VOG

Eruptions in Hawaii are gentle, with lava oozing out continually. Yet these affect the weather. The lava tubes belch out fumes containing smelly sulphur dioxide. The gas combines with moisture over the sea to create a volcanic fog or 'vog' of sulphuric acid!

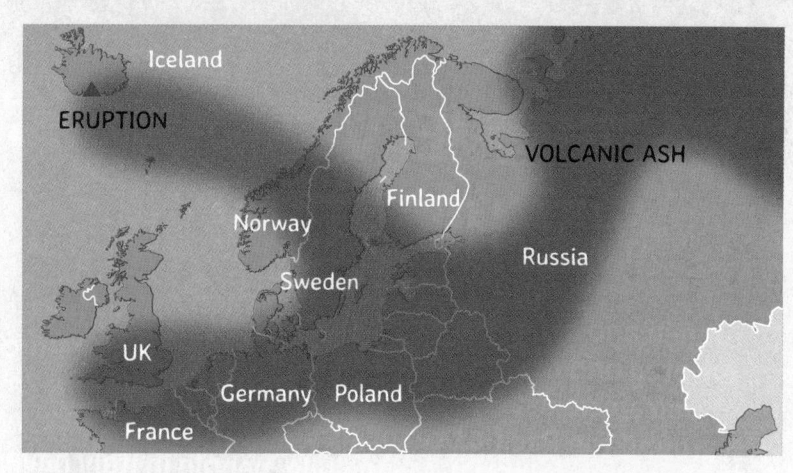

ASH MENACE

When Eyjafjallajökull volcano in Iceland erupted in 2010, it threw out a vast cloud of ash that was blown far over Western Europe. It was feared that the ash might have a catastrophic effect on the engines and electronics of jet planes. All planes were grounded for five days, and 95,000 flights were cancelled.

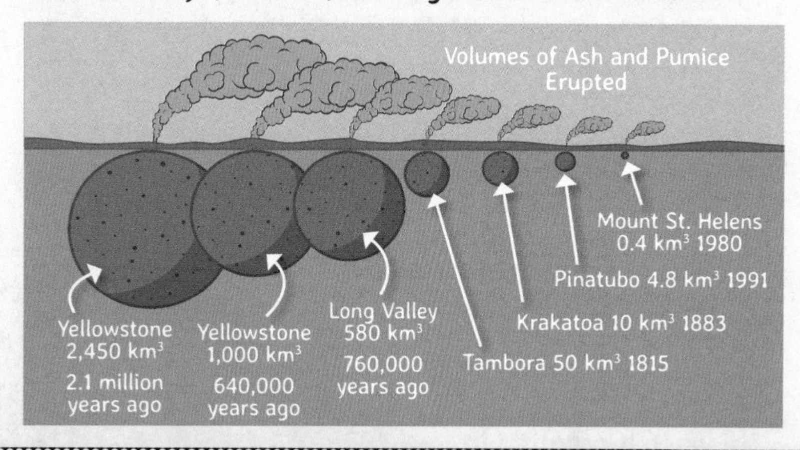

A NARROW ESCAPE

On 24 June 1982, British Airways Flight 9 flew through ash from the eruption of Mount Galunggung in Indonesia. The captain calmly announced, "All four engines have stopped. We are doing our damnedest to get them under control..." Fortunately, the engines restarted once the plane glided out of the ash cloud, and it landed safely!

EXTREME VOLCANOES IN HISTORY

Again and again throughout history, volcanoes have erupted and caused devastation. Many of the gigantic volcanic blasts have occurred in the 'Ring of Fire' around the Pacific Ocean, including Tambora (1815) and Krakatoa (1883). Meanwhile, people in Italy have been faced with the dangers of volcanoes such as Vesuvius since ancient times.

UNBELIEVABLE!
There was a volcano located where the Greek island of Santorini is now. It erupted sometime between 1645 and 1500 BC, in one of the biggest explosions ever witnessed.

TAMBORA 1815

The explosion of Mount Tambora on Sumbawa island, Indonesia, in April 1815 was the largest ever in recorded history. The noise was heard in Sumatra more than 1,930 km away. Ash clouds from the eruption spread worldwide and stopped summer coming to the northern hemisphere. About 10,000 people died because of the eruption and the following tsunamis. Around 82,000 more died because of famine and disease it resulted in.

KRAKATOA 1883

Krakatoa is a volcanic island in the Sunda Strait in Indonesia. Its eruption on 26 August 1883 was one of the deadliest in recent history, killing nearly 40,000 people. It ejected huge amounts of rock, ash and pumice and was heard thousands of kilometres away. The explosion set off a tsunami, with waves up to 40 m high, that reached Arabia about 11,000 km away.

FIERY RING

Many of the world's most dangerous volcanoes form a 'Ring of Fire' around the Pacific Ocean, where tectonic plates – the giant slabs of rock that make up the Earth's crust – crunch together. Here, magma forces its way up through thick rock, making it super-sticky. Super-sticky magma frequently clogs up the volcanic vent so that pressure builds up until the clog bursts explosively.

VESUVIUS 79 AD

In 79 AD, Mount Vesuvius, near the modern city of Naples, erupted and buried the Roman towns of Pompei and Herculaneum in rock and dust, killing thousands. The ashfall preserved much of the towns, as well as the body shapes of victims, as if in a time capsule. Some think Vesuvius is the most dangerous volcano today, since three million people live nearby.

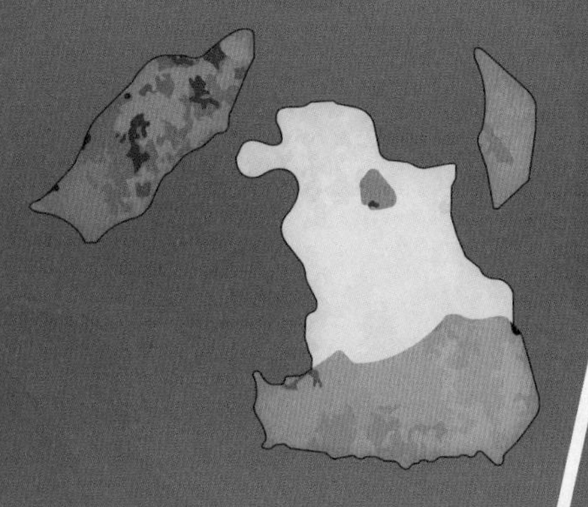

The island of Krakatoa was almost all destroyed in the 1883 eruption (right), but new eruptions beginning in December 1927 built the Anak Krakatau (Child of Krakatoa) in the same place.

I WAS THERE

Chances Peak is a volcano in the Soufrière Hills on the Caribbean island of Montserrat. Before 1995, it had been dormant for over 300 years. But that year it started to rumble and spew out dust and ash. By 1997, it was erupting continuously, setting off pyroclastic flows and lahars so devastating that people had to be evacuated.

EYEWITNESS

Eyewitness Roy Daley saw one of the pyroclastic flows coming – and was lucky to escape. "I saw the surges coming back up the hill from the pyroclastic flows, which moved at incredible speed... breaking over the walls at Bramble [village] and rushing down towards Spanish Point [on the coast]."

PLYMOUTH DESTROYED

Montserrat's capital, Plymouth, was buried under layers of ash and mud. Many homes and buildings across the southern half of the island were lost, including the only hospital and the airport.

GETTING OUT

Evacuating all the island's people in time was a hazardous task, especially since the airport was destroyed. The British warship HMS *Liverpool* played a key role in taking people off the island. Some went to the neighbouring islands of Antigua and Barbuda. Many went all the way to the UK.

DOME VOLCANO

The lava from the Montserrat volcano is very sticky and cannot easily flow away from the vent. So instead it has piled up into a large, dome-shaped mass. The eruptions frequently blast solidified magma out as clouds of ash and pyroclastic flows.

SCORCHING FLOWS

The most dangerous features of the Montserrat eruptions were the pyroclastic flows: racing streams of scorching ash that incinerated everything in their path. People who stayed on to try and protect their homes believed they could guess when the flows were coming and make it to safety. Sadly, they were not always right.

WHY LIVE IN THE WAY?

Today, nearly half a billion people live on or close to active volcanoes. There are even major cities close to active volcanoes. For example, Mexico City is near Popocatépetl, and Naples near Vesuvius in Italy. They are dangerous places to live. But for the people who live there, the benefits outweigh the dangers.

UNBELIEVABLE!

Etna is the world's most active volcano, throwing out more lava than any other. It even gave us the name 'volcano', for the Ancient Romans believed it was the home of the fire god Vulcan.

IN THE VOLCANO'S CRADLE

In Kamchatka in the far east of Russia, 180,000 people live in the port city of Petropavlovsk-Kamchatskiy – completely surrounded by active volcanoes. The volcanoes are dangerous though very beautiful, but the port gives access to the Pacific Ocean for ships and for fishing, making the risks worthwhile for the inhabitants.

INTO HELL

Miners work in the crater of East Java's Ijen volcano, collecting sulphur to make matches and fertilizer, and vulcanising rubber for car tyres. Even if there is no eruption, the water in the crater is acidic enough to dissolve clothes, eat through metal and damage lungs!

THE WILDEBEEST NURSERY

Ol Doinyo Lengai means 'Mountain of God' in the language of the local Maasai people of Kenya. It is an active volcano that throws out a unique 'carbonatite' ash that makes the soil around the volcano very good for grass. The rich pastures are the reason that wildebeest stop here to give birth to calves on their long annual migration. The grasslands also attract Maasai herdsmen.

VINTAGE VOLCANO

Many wine growers say wine grown on volcanic soil is the best. Volcanic ash makes light, well-drained soil, rich in nutrients such as magnesium, calcium, sodium, iron and potassium. So Sicilian wine growers risk the dangers to grow grapes on the slopes of Mount Etna.

ONE STEP AHEAD OF THE VOLCANO

Volcano experts are called volcanologists, and it is their task to learn about volcanoes and detect signs ahead of an eruption to warn people living nearby. They look for clues such as pressure building up in the magma chamber beneath a volcano, or unusual gases being released. But monitoring a volcano can be a dangerous business.

UNBELIEVABLE!

A change in the chemistry of lava could warn of a coming eruption. So volcanologists wear protective clothing to take samples from a fresh, hot lava stream.

DRONES IN ACTION

Robot flying machines called drones are becoming important weapons in the volcanologists' armoury. They can fly right inside a volcano's crater where no human would dare set foot. Here, they may detect the clearest possible signs that a volcano is likely to erupt.

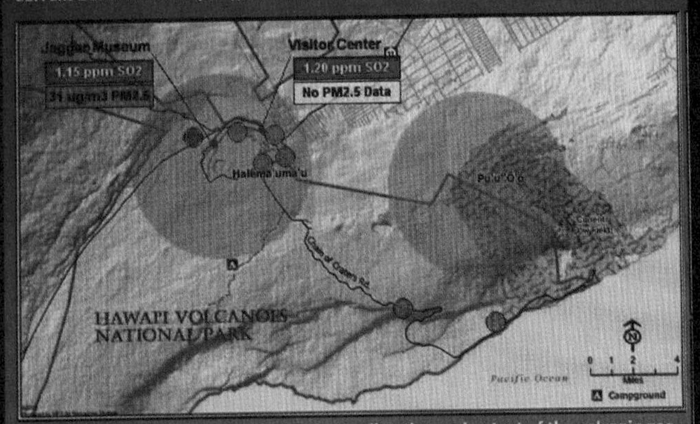

Hawai`i Volcanoes National Park
Current Sulfur Dioxide (SO_2) and Particulate ($PM_{2.5}$) Conditions
Visitor Center
1.15 ppm SO2
1.20 ppm SO2
No PM2.5 Data
Halema'uma'u
Pu'u'O'o
Chain of Craters Rd.
HAWAI'I VOLCANOES NATIONAL PARK
0 1 2 4
Pacific Ocean
Campground
The map above indicates the approximate direction and extent of the volcanic gas plumes (yellow wedges) from Halema'uma'u and Pu'u O'o. Within the plumes are unhealthy concentrations of sulfur dioxide (SO_2) and small particles ($PM_{2.5}$). Color-coded health advisories (color circles or label boxes) are issued when the plumes affect surface areas.
ulfur dioxide gas plumes cross roads near Halema`uma`u and low on
hain of Craters Road. Sensitive individuals should limit exposure in these

LOOKING FOR GAS

When a volcano is about to erupt, magma moves up the vent and begins to give off gases such as sulphur dioxide and hydrogen sulphide. If volcanologists can get close enough, they might be able to detect rises in these gases in the smoke billowing from side chimneys called fumaroles, or in the chemistry of crater lakes.

GROUND SHIFT

A telltale sign of magma pressure building up under a volcano could be a slight change in the volcano's shape. Meters are placed in the ground on a volcano and with the aid of GPS satellites they can detect changes of just a few millimetres in the shape of the ground. Even a small change like this might indicate an imminent eruption.

GROUND SHAKING

For many volcanologists, the key clue to a coming eruption is 'seismic activity'. By this they mean the ground on the volcano shaking and rumbling a little. Almost all the world's major volcanoes are now continually monitored for any shaking, using seismometers either slightly away from the volcano or right on its slopes.

COMPARING ERUPTIONS

Volcanologists compare the scale of eruptions using the Volcanic Explosive Index (VEI). It uses things such as volume of ash and cloud height to indicate just how explosive an eruption was, from gentle (0) to mega-colossal (8).

UNSEEN EXTREME VOLCANOES

Many volcanoes, like Mount Fuji or Kilimanjaro, are hard to miss, towering high above their surroundings. But many others are almost entirely invisible, erupting out of sight under the sea. Indeed, there are thought to be a million or more volcanoes below sea level, producing more than three-quarters of all the magma erupted on the Earth.

EYEWITNESS

On 14 November 1963, crew of the fishing boat Ísleifur II spotted dark smoke over the sea south of Iceland. It was not a boat on fire, as they thought, but the new island of Surtsey being born. In a few days, it had reached over 500 m long and 45 m high.

The Mid-Atlantic Ridge can be seen briefly above the sea in Iceland's Thingvellir gorge

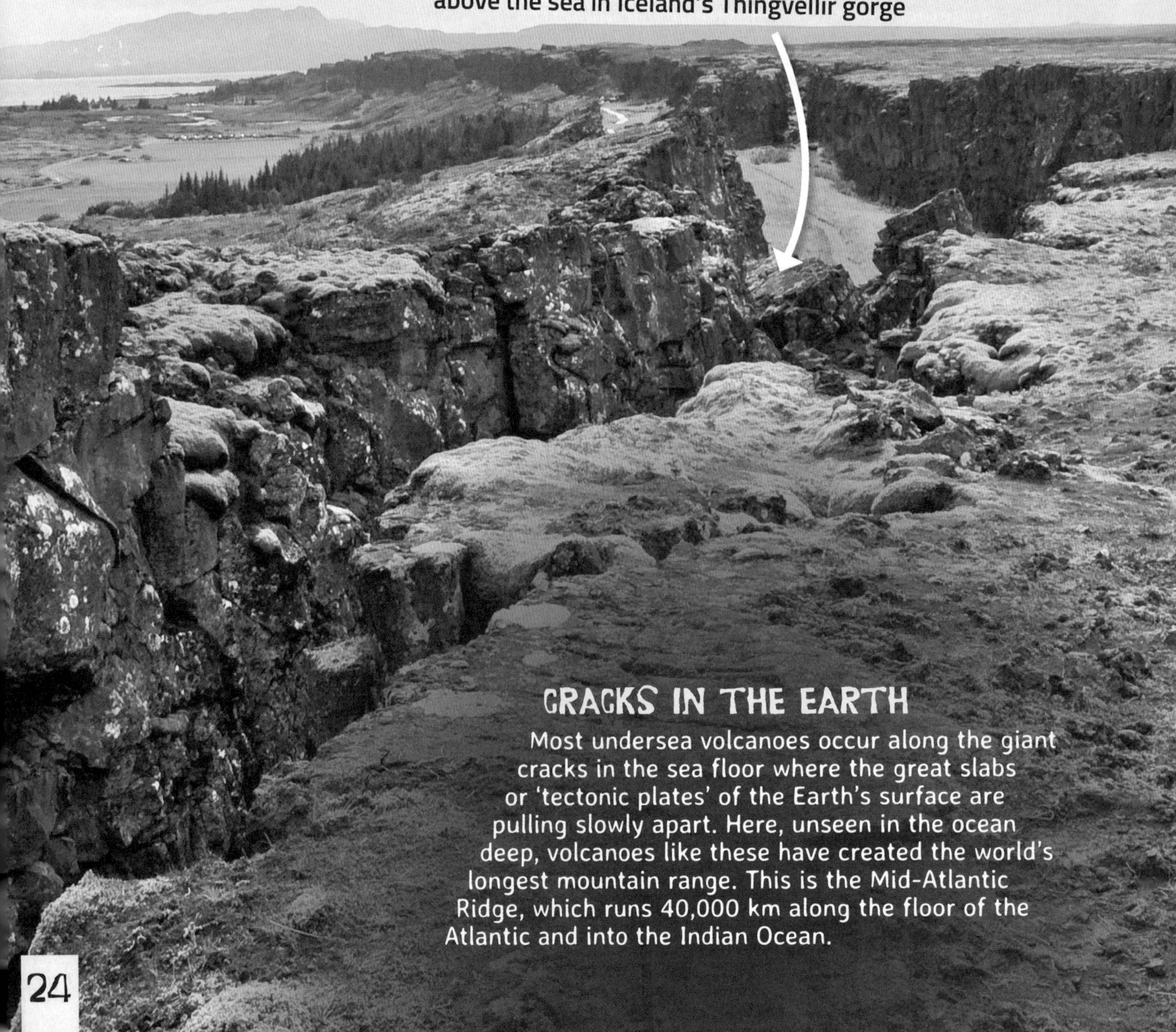

CRACKS IN THE EARTH

Most undersea volcanoes occur along the giant cracks in the sea floor where the great slabs or 'tectonic plates' of the Earth's surface are pulling slowly apart. Here, unseen in the ocean deep, volcanoes like these have created the world's longest mountain range. This is the Mid-Atlantic Ridge, which runs 40,000 km along the floor of the Atlantic and into the Indian Ocean.

BLACK SMOKERS

Geysers like Yellowstone's Old Faithful are chimneys where bursts of superhot water whoosh out after it is heated under pressure by volcanic activity underground. Amazingly, there are geysers, or 'hydrothermal vents', under the sea, too. Only they are called 'black smokers' because the jets of water are mixed in with dark clouds of minerals.

Black smoker

Water seeping through ocean floor

Super-heated water

Heat from magma below

NEW ISLANDS

New islands can appear in the sea overnight, when an undersea volcano grows high enough to erupt above the surface. This happened at Hunga Tonga in the Pacific in 2009, Nishinoshima in Japan in 1973 and, even more spectacularly, with the island of Surtsey off Iceland in 1963.

HOT GIANT

Hawaii's Mauna Kea is 10,000 m tall, taller than Qomolangma's 8,844 m! But most of it is underwater. Mauna Kea is a 'hot spot' volcano. 'Hot spots' are rare places where fountains of hot molten rock rise up through the Earth's interior and scorch their way out through the crust.

EXTREME ERUPTIONS

Every now and then in Earth's history, there have been volcanic events so mighty that they have created worldwide cataclysms. Some 252 million years ago, 96% of all sea creatures and 70% of all land creatures may have been wiped out when volcanic eruptions poisoned the air and caused massive global warming. Could such a terrible event happen again?

THE WORLD'S GREATEST DISASTER

Scientists call the terrible event that wiped out so much life on Earth the 'Permian-Triassic Mass Extinction', or more simply the 'Great Dying'. Some think the damage was done by a giant meteorite smashing into the Earth. But most think it was triggered by a continuous series of massive eruptions in Siberia. These spread 2,000,000 km^2 of lava over Siberia, called the Siberian Traps. The eruptions also released gases and ash that turned rain acid and trapped so much of the sun's heat in the atmosphere that the world scorched.

COULD IT HAPPEN AGAIN?

It is certain that there will be a cataclysmic 'supervolcano' eruption like Yellowstone (below) in the future. And there is no way to prevent it. But scientists are trying to discover ways of monitoring the pressure of underground magma to predict whether one is imminent. Most experts agree that no supervolcano is likely to erupt in the foreseeable future. But the foreseeable future spans just 10 years...

HUMAN ESCAPE?

Around 74,000 years ago, one of the largest ever volcanic eruptions, the eruption of Toba in Indonesia, flung huge amounts of ash and dust into the air. As all this material spread around the world, it blocked out the sun and made the world turn icy cold. It's thought that only a handful of humans – just a few thousand – made it through this terrible time. That may be why today we all have the same few ancestors from back then.

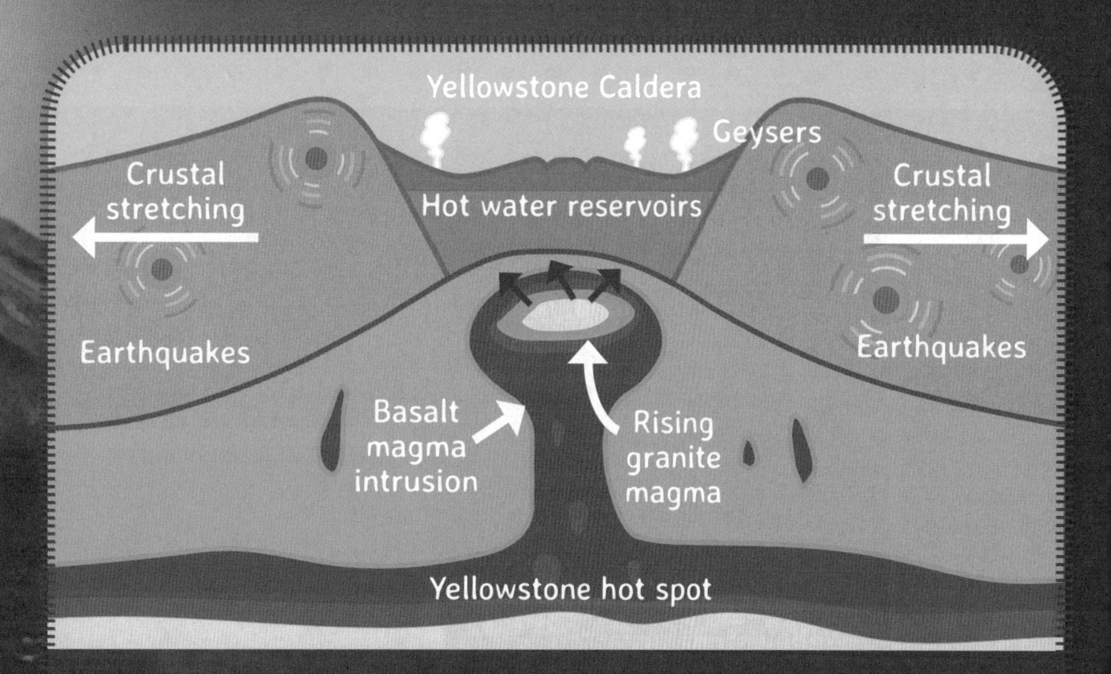

YELLOWSTONE SUPERVOLCANO

Yellowstone is actually one of the world's biggest volcanoes. It is what volcanologists call a 'supervolcano'. It last erupted 640,000 years ago, flooding the region with enough lava to fill up the Grand Canyon. Yellowstone looks fairly mild today, but it is now overdue for another eruption. Scientists think that if it did erupt, the USA would be covered in ash more than 1 m deep.

TIMELINE

65 million years ago
A massive lava eruption in what is now Southern India created the Deccan Traps, a thick lava plateau twice the area of Texas

1645-1500 BC
The island volcano of Thera in the Mediterranean erupted in one of the largest explosions

79 AD
The eruption of Mount Vesuvius in Italy smothered the Roman towns of Pompei and Herculaneum in thick ash

1600
The biggest eruption in American history, of Huaynaputina in Peru, destroyed the city of Arequipa and created a cloud of ash that affected weather worldwide and caused a famine in Russia

1815
The eruption of Tambora in Indonesia was the biggest in recent history. The vast amounts of ash thrown up caused a cool summer around the world

65 mya

27,000 years ago
A supervolcano eruption on North Island, New Zealand, created the enormous crater now filled by Lake Taupo

1883
Krakatoa in the Sunda Strait, Indonesia, erupted with a bang heard thousands of kilometres away and triggered a massive tsunami

640,000 years ago
A supervolcano erupted (again) to create the Yellowstone area

1783
The eruption of Laki in Iceland spread poisonous gases across Western Europe

1 million years ago
Ascension Island in the Atlantic was created by a volcanic eruption

1902

On the French island of Martinique in the Caribbean, a pyroclastic flow from the eruption of Mount Pelée killed some 29,000 people in the town of St. Pierre

1980

Mount St. Helens in Washington State erupted with the most devastating eruption in US history, killing 57 people and doing huge damage

1963

The Icelandic island of Surtsey was created by an undersea volcanic eruption

1985

A mudslide triggered by the eruption of Colombia's Nevado del Ruiz volcano killed 25,000 people

1991

The eruption of Pinatubo in the Philippines ejected more than 4 km^3 of material into the air and sent up a column of ash 35 km high

1982

El Chichón began to erupt in Mexico

2010

Dust from the eruption of Eyjafjallajökull in Iceland downed all air flights in Western Europe.

Mount Merapi in Java erupted, forcing the evacuation of 350,000 people

1983

Kilauea began to erupt in Hawaii

1912

Novarupta in Alaska was the biggest eruption of the 20th century and drained magma from nearby Mount Katmai, making it collapse

1986

A cloud of carbon dioxide bubbling from under Lake Nyos in Cameroon suffocated 1,746 people

HOT AND EXPLOSIVE

Amazing facts about volcanoes

HIGH PEAK

The furthest point from the centre of the Earth is not the peak of Qomolangma, but the peak of the volcano Chimborazo in Ecuador. This is because the Earth spins very fast and so it bulges in the middle. This makes it fatter at the equator, and Chimborazo is very near the equator.

BIG MONS

There are some big volcanoes on Earth, but the biggest in the solar system is on Mars. Its name is Olympus Mons and it is gigantic: 600 km wide and 21 km high.

IO

Io, Jupiter's moon (shown above, in front of its large parent) is the most volcanic place known. The whole surface of this moon is a cauldron of volcanic activity. In fact, its surface is constantly changing through volcanic eruptions.

LIGHT ROCK

Pumice is a very light and porous rock that forms when the glassy froth of lava cools and solidifies. It is the rock that actually floats!

HAPPENING NOW

There are at least 20 volcanoes erupting around the world at this very moment! Between 50 and 70 volcanoes erupt every year, and over the last decade about 160 have been active. Scientists estimate that in the last 10,000 years, 1,300 volcanoes erupted.

VOLCANIC HATCHERY

On the Indonesian island of Sulawesi, maleo birds don't sit on their eggs to keep them warm. They cleverly use volcanic heat to help hatch their eggs! They bury the eggs in soil or sand near volcanoes. When the chicks hatch, they claw their way up to the surface.

HOT LAVA

Lava might not always move very fast, but you don't want to stand in its way. Its temperature can reach 1,250°C and will burn through pretty much everything!

VOLCANIC WORLD

We owe much of our world to volcanoes. More than 80 per cent of Earth's crust was made by volcanoes. Although a lot is now covered with a veneer of sediments, most sediments are fragments of crumbled volcanic rock. The sea floor and some mountains were formed by countless volcanic eruptions. Gases from volcanoes also formed Earth's atmosphere.

WORST VOLCANIC ERUPTIONS

LARGEST EXPLOSIVE ERUPTION EVER

Guarapuava-Tamarana-Sarusas,
Paraná Traps, Brazil
132 million years ago (mya)
Volume: 8,600 km³ — enough to cover New York State to a depth of over 60 m

LARGEST IN THE LAST 25 MILLION YEARS

Lake Toba, Sunda Arc, Indonesia
74,000 years ago
Volume: 2,800 km³ — enough to cover New York State to a depth of over 20 m

LARGEST IN THE LAST 50,000 YEARS

Taupo, New Zealand
27,000 years ago
Volume: 1,170 km³— enough to cover New York State to a depth of over 8 m

LARGEST ERUPTION OF LAVA

Mahabaleshwar-Rajahmundry Traps,
Deccan Traps, India
65 mya
Volume: 9,300 km³ — enough to cover New York State to a depth of over 70 m

DEADLIEST ERUPTION

Mount Tambora, Indonesia, 1815
Cost in lives: 92,000

DEADLIEST IN THE LAST CENTURY

Mount Pelée, Martinique, 1902
Cost in lives: 29,000

ONE OF THE DEADLIEST IN THE LAST 50 YEARS

Mount Pinatubo, Philippines, 1991
Cost in lives: 300

DEADLIEST VOLCANIC EFFECTS

Huaynaputina, Peru, 1600
Cost in lives: 2 million Russians died of famine after the huge ash cloud caused extremely cold weather across the world

INDEX

THE AUTHOR

John Farndon is Royal Literary Fellow at City&Guilds in London, UK, and the author of a huge number of books for adults and children on science, technology and nature, including such international best-sellers as *Do You Think You're Clever?* and *Do Not Open.* He has been shortlisted six times for the Royal Society's Young People's Book Prize for a science book, with titles such as *How the Earth Works, What Happens When...?* and *Project Body* (2016).

Picture Credits (abbreviations: t = top; b = bottom; c = centre; l = left; r = right)
© www.shutterstock.com:
1 c, 2 l , 3 tl, 4 c, 6-7 c, 9 tr, 11 cr, 11 br, 12 c, 12 br, 13 tl, 13 tc, 13 tr, 13 br, 14 c, 15 tl, 17 cr, 17 cl, 20 c, 21 tl, 21 tr, 21 br, 23 bl, 24 c, 25 bl, 28 br, 29 br, 29 tr, 30 c, 30 bl,

NASA:
19 tr,

9, br = Universal Images Group North America LLC / Alamy Stock Photo. 11, tr = PhotoAlto sas / Alamy Stock Photo. 11, cr = FLPA / Alamy Stock Photo. 18, c = Barry Lewis / Alamy Stock Photo. 18, br = Barry Lewis / Alamy Stock Photo. 19, tr = NASA. 19, br = Colin Burn-Murdoch / Alamy Stock Photo. 22, c = ARCTIC IMAGES / Alamy Stock Photo. 22, bl = imageBROKER / Alamy Stock Photo. 23, t = Craig Buchanan / Alamy Stock Photo. 23, br = ARCTIC IMAGES / Alamy Stock Photo. 25, cl = epa european pressphoto agency b.v. / Alamy Stock Photo. 28, tr = Stocktrek Images, Inc. / Alamy Stock Photo. 31, tr = Kevin Schafer / Alamy Stock Photo

极端自然灾害

火山

约翰·法恩登（英）著
陈泰之 译

外语教学与研究出版社
FOREIGN LANGUAGE TEACHING AND RESEARCH PRESS
北京 BEIJING

我们的地球烦躁不安，脾气暴躁，随时可能在许多地方突然发生剧烈的火山喷发。事实上，此时至少有20座火山正在喷发之中！火山喷发出的火山灰能够掩埋整座城市，喷发能引起灼热的崩流，毒化空气，引发海啸，并极大地改变我们的气候。

了解一下火山已经造成的最严重的危害，以及将来潜在的危害。看看它们是如何发生的，“热点”在哪里，为什么人们还要居住在它们附近，那些聪明的火山学家为了预测未来的火山喷发在做些什么……

目录

火山

在地球表面的某处，熔化的岩石冲破坚固的地壳，从地球炽热的内部喷出，这样就形成了火山。有时，熔岩以岩浆形式从裂缝中流出，形成火山岩（即流出地面的熔岩）。而有时，熔岩会连续喷发形成锥形火山。火山可能会突然急剧爆发，并在爆发过程中释放出大量高温物质。

火山是怎样喷发的？

在大部分火山下面都有一个岩浆房，这里贮存着高温岩浆。随着时间的流逝，下面的岩浆堆积得越来越多，岩浆房里的压力也越来越大。最终，当压力增加到足够大的时候，岩浆就会通过火山的狭窄通道（又称“火山口”）冲出地表。

目击者

公元79年，意大利的维苏威火山发生可怕的喷发，普里尼型喷发由此得名。罗马学者老普里尼死于那场火山喷发，他的外甥在海上的一艘船上目睹了这一切。

在大部分剧烈的喷发中经常出现静电积聚造成的闪电

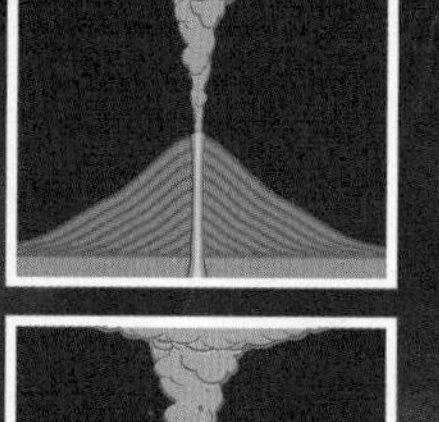

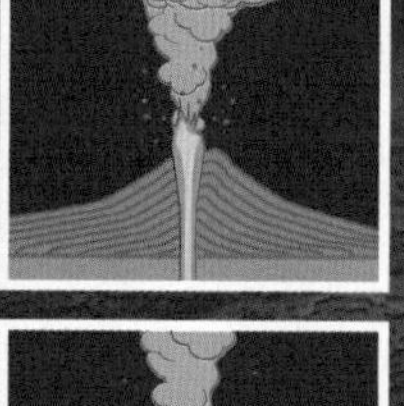

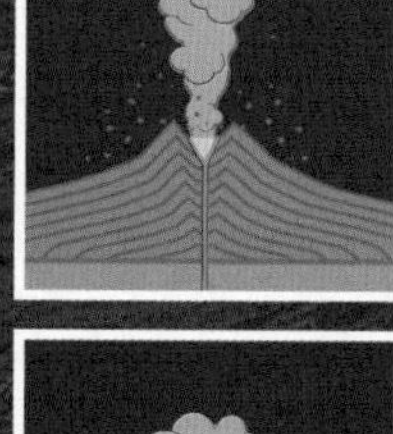

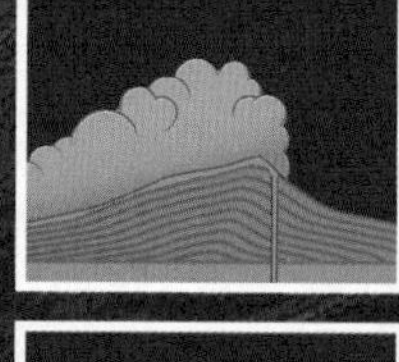

称为“火山弹”的大块物质倾泻而下，散落在火山四周

火山喷发时将火山通道中的大量堵塞物炸开，形成大量的火山灰和岩屑

有时，岩浆从火山侧面喷出，形成较小的“次级锥体”

主通道

一旦火山灰与岩屑从通道中被炸出来，熔化的地下岩浆就会奔涌而出，并以熔岩流的形态顺山而下

喷发时刻！

在喷发的火山内部，上一次火山喷发产生的物质凝固，并把主通道堵住。当上升的岩浆冲破堵塞物的时候，新一次的喷发就开始了。随着岩浆喷出，火山内部压力开始降低。二氧化碳气体和水蒸气形成的气泡快速形成，推动岩浆飞速喷出，就像汽水从摇晃后的瓶子中喷出一样。

岩浆房，火山喷发之前岩浆在火山底部积聚

快速而剧烈

普里尼型火山喷发具有超强爆炸力，源源不断地喷发出大批气体和火山灰，伴随着被称为“浮石”的泡状凝固的热岩浆碎屑物。这种喷发的持续时间很少能超过几周，有时候连一天都不到。

火山喷发的类型

这些只是火山喷发的部分类型：

普里尼型喷发最剧烈，产生高耸入云的火山气体和火山灰，它们呈现出平顶云团的形状，扩散到远方。

乌尔坎诺型喷发产生的气体爆炸程度稍弱一些，和火山灰混合在一起，形成黑色翻滚的云团。云团卷绕在一起，快速升起。

斯特朗博利型喷发爆发快，喷发物中混杂着火山灰和炽热发光的熔岩碎块。

培雷型喷发会抛出高速前进的由炽热火山灰和岩屑组成的崩流，人们称之为火山碎屑流。

夏威夷型喷发的剧烈程度不高，但却连绵不断。它喷发出大量流淌的熔岩，这些熔岩能流动很远，然后冷却并固化，形成坡度小的盾状火山。（见第10页）

大爆炸

1991年6月15日，菲律宾的皮纳图博火山发生了爆炸性喷发，这是近期发生的最大的火山喷发之一。它喷射出高速前进的由炙热的火山灰与火山气体组成的崩流（火山碎屑流），以及浩大的泥流（火山泥流），形成了直径为400千米的火山灰云团。

极端火山的危害

火山喷发会带来各种各样的危害。它们喷射出的气团和火山灰令人窒息。它们喷发出的炽热岩屑会轰炸周围地区；它们不停顿地喷发出通红的熔岩流和炽热的崩流，摧毁挡在途中的一切事物；它们甚至还能够诱发地震和滑坡。

目击者

约瑟夫·恩克温经历了1986年喀麦隆的尼奥斯湖灾难（见第11页右下角图片）。“我说不出话，失去了知觉。我张不开嘴，因为当时我闻到了可怕的气味……”

火山的类型

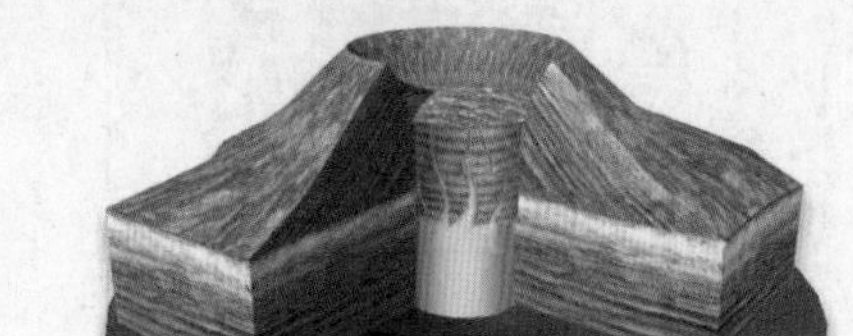

破火山口组合体中有一个从山上沉入岩浆房的火山口，就像黄石公园里的火山那样。

穹状火山因黏稠的熔岩汩汩地流出，形成穹顶状的山丘而得名。法国的多姆山就是一座穹状的死火山。

裂隙式火山有很长的地面裂缝，熔岩沿裂缝连续溢出，这种火山的爆炸性很小，冰岛的许多火山即为裂隙式火山。

盾状火山因熔岩流淌具有宽广性而形成。夏威夷的冒纳凯阿火山就是盾状火山。

层状火山又叫复合型火山，如意大利西西里的埃特纳火山，由连续喷发而出的多层熔岩和火山灰构成。

灰渣锥型火山是由火山灰和火山渣形成的锥状火山。位于印度尼西亚的喀拉喀托火山为灰渣锥型火山。

泥石流

一些火山喷发会引起熔岩、泥浆和水的破坏性流动，即火山泥流。当火山喷发产生的热量融化了积雪和冻土时，火山泥流常常随之发生。泥浆疾流下山，流速快如车速。1985年，哥伦比亚鲁伊斯火山喷发后，火山泥流袭击了阿尔梅罗镇，造成25,000人死亡。

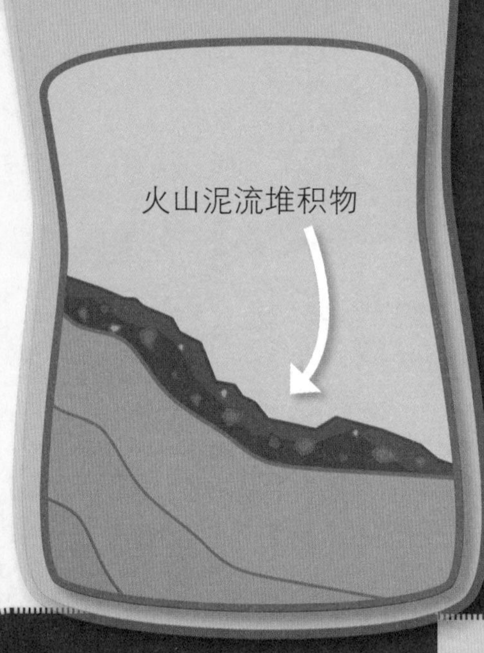

火山的“军械库”

火山能引发多种危险！

• 熔岩流——流淌着的炽热的熔岩——十分恐怖又势不可挡，好在熔岩流通常速度较慢，给人们足够的时间逃离。但熔岩流会摧毁建筑物。

• 灼热、令人窒息的火山灰和火山渣组成的云团，可以覆盖火山之外很远的地方，危害巨大，人们几乎没有逃离的机会。火山灰的重量能够压塌屋顶。

• 来自于火山岩屑的纤细粉尘逐渐飘落，慢慢造成人和动物的窒息。

• 火山碎屑流指的是由火山灰、火山渣和火山气体组成的高温、高速崩流。

毒气中毒！

最怪异的火山灾害之一发生于1986年8月21日的喀麦隆，仅在几分钟内，1,700多人和许多动物死于无形之中。原因是附近的尼奥斯湖底发生了轻微的火山喷发，喷发释放出大量的二氧化碳，充满整个湖面。然后该气体由湖面释放至空气中，导致人和动物窒息而亡。

最危险的火山

碰到火山喷发是一件危险的事。然而许多人住在很久没有喷发的火山附近，似乎很安全。但是，这些火山是真的“死了”（彻底不再喷发），还是“休眠了”（暂时不喷发），或者是在等待下一次的毁灭性喷发？

骇人的喷发

之前很多人认为华盛顿州的圣海伦斯火山是座死火山，因为100多年来它都无声无息。然而，1980年3月20日圣海伦斯火山突然喷发，成为美国历史上最致命、损失最惨重的一次火山喷发，造成57人死亡。

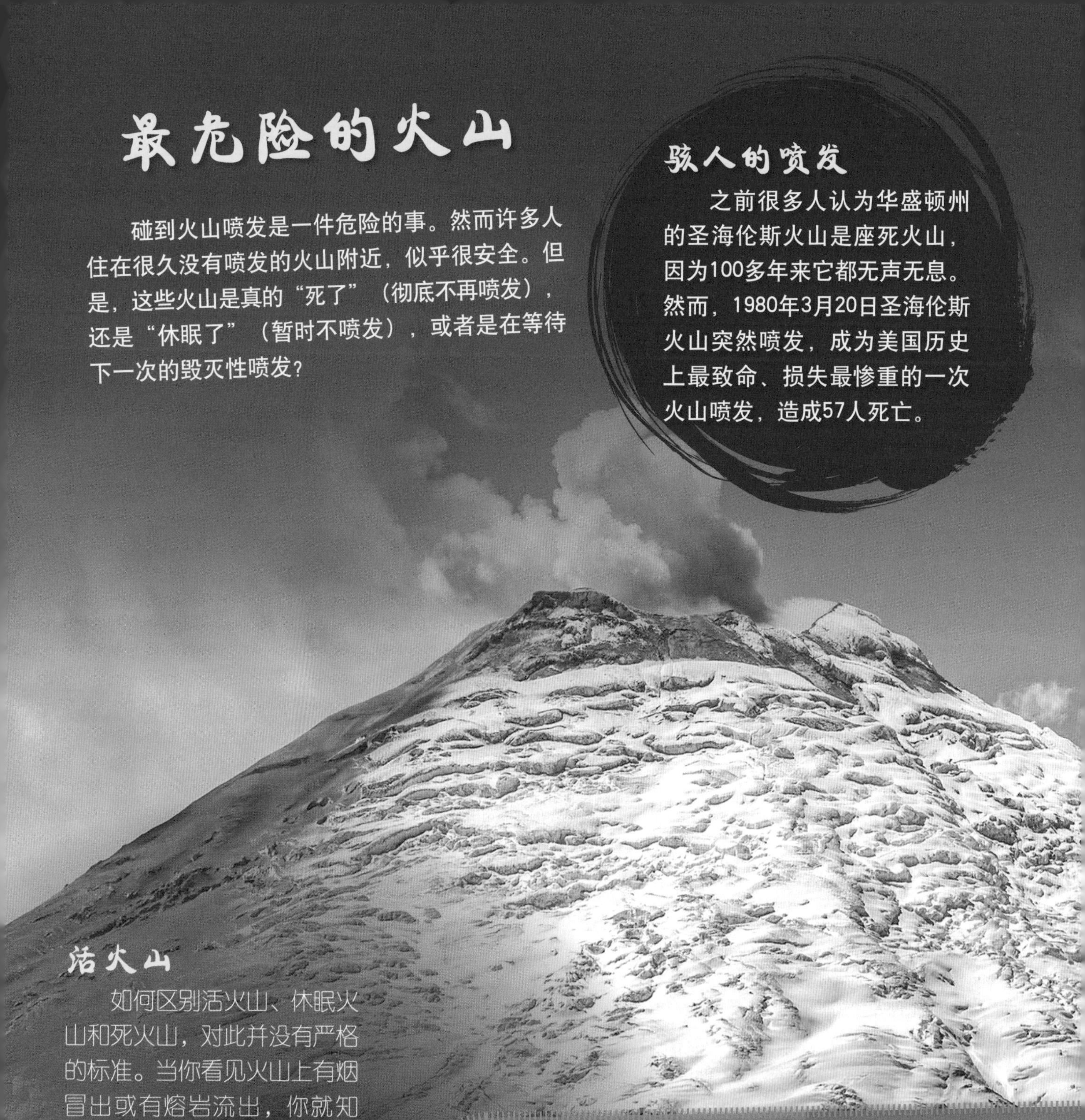

活火山

如何区别活火山、休眠火山和死火山，对此并没有严格的标准。当你看见火山上有烟冒出或有熔岩流出，你就知道它是活的，就像美国的基拉韦亚火山、日本的樱岛火山，以及厄瓜多尔的科托帕希火山（见上图）。火山科学家通常认为，某座火山如果在近一万年内喷发过，那么它就是活火山。

休眠火山

休眠火山是指已经一万年没有喷发，但仍有可能在未来某一天喷发的火山。非洲的乞力马扎罗火山已经有36万年没有喷发了，但它还没完全变成一座死火山。

死火山

死火山是已经数万年、数十万年，甚至数百万年没有喷发的火山。它们似乎不大可能再喷发，因为山底的岩浆已经冷却、固化。众所周知，苏格兰的爱丁堡城堡就修建在一座死火山之上。

活动还是休眠？

日本的富士山是世界上最著名的火山之一。它已经300多年没有喷发了。那么，它是正在休眠呢，还是在准备着一次突然喷发？即使今天，这座山上的小洞——喷气孔——也还释放着烟雾。没有人可以确定这些烟雾是不是危险的信号。

睡巨人

华盛顿州的雷尼尔山白雪皑皑，看起来宁静而美丽。自1894年以来它就没有喷发过，如此看来，它也许是休眠了。但是它附近有许多居民，如果它突然喷发，所有冰雪都会在可怕的雪崩中融化。

极端气候效应

火山不只对地面和地貌造成影响，它们也能改变大气层。如果发生爆炸性的火山喷发，它会向大气层抛出大量火山灰和污染性气体，不仅能够改变当地气候，也能改变全球气候。

你知道吗？

1600年2月19日，秘鲁埃纳普蒂纳火山喷发，造成远在俄国的200万人死亡。火山抛射出的火山灰使俄国那一年的夏天十分寒冷，造成了可怕的饥荒。

火山风暴

如果火山喷发抛射出的碎屑和熔岩不够把你吓得够呛，那么试试它经常带来的猛烈风暴！炸到高空的灰烬颗粒吸附在水滴上，变成了超级强烈的雷雨云，带来大量降雨，并伴随电闪雷鸣。

讨厌的火山烟雾

夏威夷的火山喷发很温和，熔岩不断地向外流出。即使如此，它们还是会影响气候。熔岩通道喷出的烟雾含有难闻的二氧化硫。二氧化硫和海上的水汽结合，形成了含有硫酸的火山烟雾。

火山灰的威胁

2010年，冰岛埃亚菲亚德拉冰盖下的火山喷发，释放出大量火山灰，这些灰被吹到西欧上空。因为担心火山灰可能给喷气式飞机的发动机和电子设备造成灾难性的影响，这些地区的所有飞机停飞五天，95,000次航班被取消。

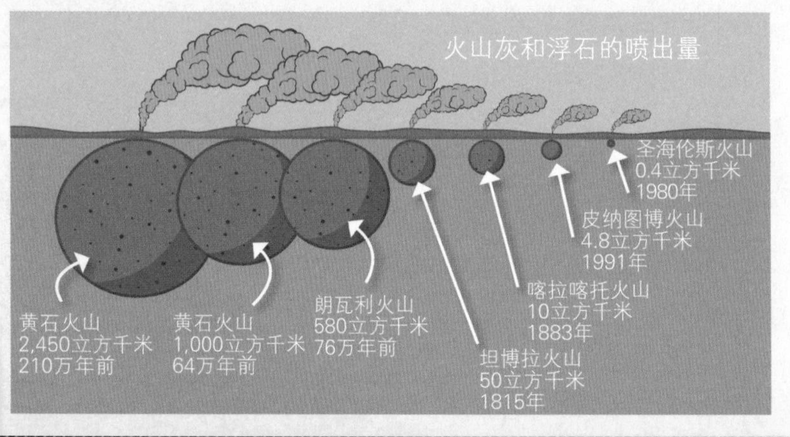

侥幸逃脱

1982年6月24日，英国航空公司9号班机飞入印度尼西亚加隆贡火山喷发产生的火山灰云。机长平静地宣布："所有四台发动机都停止了工作。我们在尽最大努力掌控它们……"幸运的是，飞机刚刚滑翔出火山灰云团，四台发动机就恢复工作了。最终，此次航班安全降落！

历史上的火山大喷发

历史上一次又一次地发生火山喷发，造成灾难。许多大的火山喷发发生在环太平洋火山带上，包括坦博拉火山喷发（1815年）和喀拉喀托火山喷发（1883年）。而意大利人自古以来就面临着维苏威等火山喷发的威胁。

你知道吗？

现在的希腊桑托林岛曾是一座火山。它在公元前1645年至公元前1500年期间发生过喷发，那次喷发是史上最大的火山喷发之一。

1815年的坦博拉火山喷发

1815年4月，印度尼西亚松巴哇岛上的坦博拉火山喷发，这场喷发是有历史记录以来最大的一场。在1,930千米之外的苏门答腊都能听到声响。喷发出的火山灰云扩散到了全世界，在北半球阻止了夏天的来临。约10,000人死于这次喷发及接踵而来的数次海啸。这次火山喷发还导致了饥荒和疾病，又夺去了约82,000人的生命。

1883年的喀拉喀托火山喷发

喀拉喀托是印度尼西亚巽他海峡中的一座火山岛。1883年8月26日它的喷发是近代历史上最致命的火山灾害之一，夺去近40,000人的生命。它喷射出大量岩石、火山灰和浮石，响声传到了数千千米以外。爆炸引发了浪高40米的海啸，海啸一度冲击到11,000千米之外的阿拉伯半岛。

环太平洋火山带

世界上许多最危险的火山组成了环太平洋火山带，地壳板块——构成地壳的巨大岩石块——在环太平洋火山带上互相碰撞挤压。岩浆从厚岩石的缝隙中挤出，变得愈发黏稠。超黏的岩浆经常堵塞火山的通道，使压力积聚，直到堵塞物在喷发中爆炸。

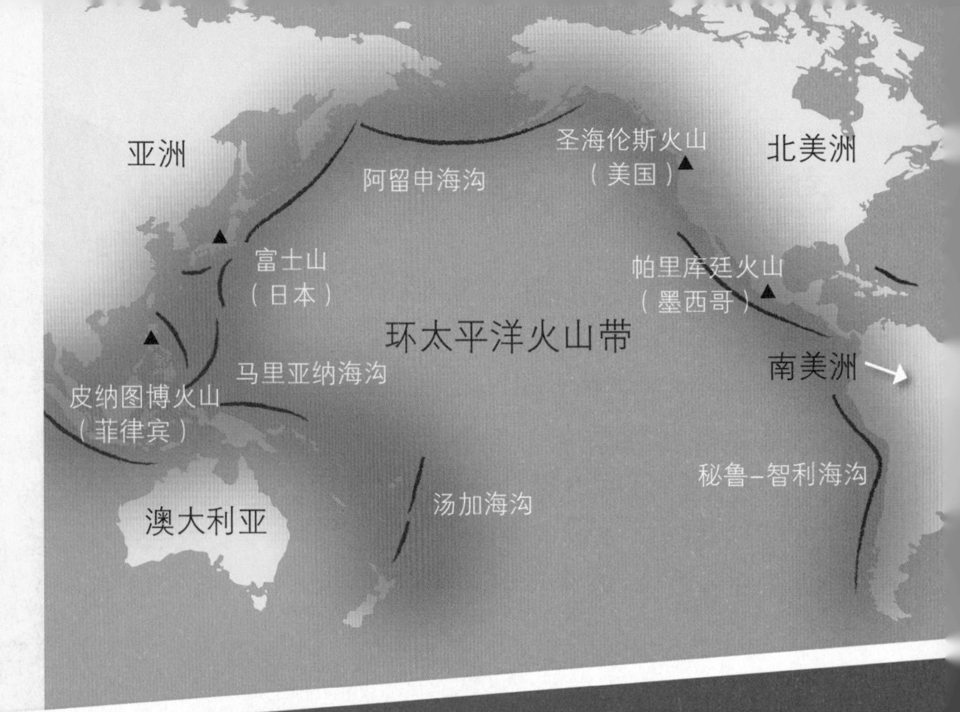

公元79年的维苏威火山喷发

公元79年，位于当今那不勒斯城附近的维苏威火山发生喷发，岩石和灰尘掩埋了罗马古城庞贝和赫库兰尼姆，导致数千人死亡。飘落的火山灰保存了古城的大部分原貌，包括受难者的遗体轮廓，就像把古城放进了时光胶囊一般。有人认为维苏威火山是当今最危险的火山，因为它附近居住了300万人。

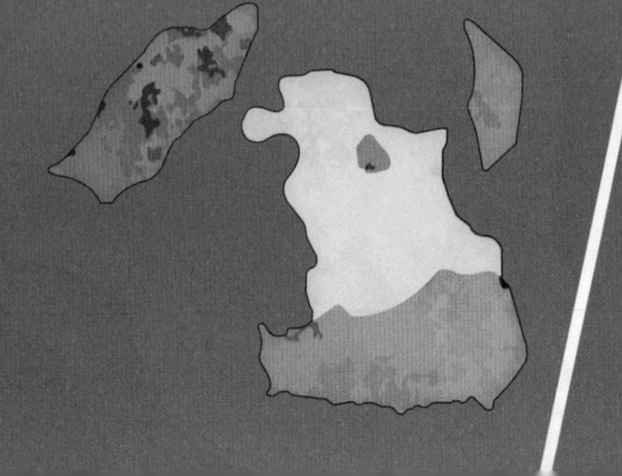

喀拉喀托岛几乎完全被1883年的火山喷发毁灭（见右图），但是，1927年12月开始的新的喷发在相同位置造出了阿纳喀拉喀托（喀拉喀托之子）火山。

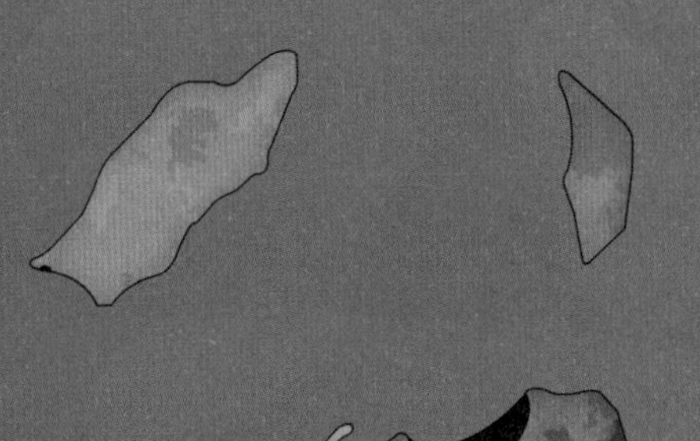

亲历火山喷发

昌色斯峰是加勒比地区蒙特塞拉特岛苏弗里耶尔山中的一座火山。1995年，在休眠了300多年以后，它开始隆隆作响，喷出火山灰和尘土。直到1997年，它都一直不停地喷发，具有强大摧毁力的火山碎屑流和火山泥流迫使人们撤离了家园。

目击者

目击者罗伊·戴利目睹了奔来的火山碎屑流，并且侥幸逃脱。“我看见火山碎屑流以惊人的速度涌上小山……捣毁布兰博尔村的院墙，冲向（岸边的）西班牙角。”戴利回忆说道。

普利茅斯被毁

蒙特塞拉特首府普利茅斯被一层又一层的火山灰和泥浆覆盖。在岛的南半部，许多房屋和建筑荡然无存，包括唯一的一家医院和一个机场。

离去

及时疏散全岛居民是一项艰巨的任务，尤其是在机场被毁之后。英国皇家海军战舰“利物浦”号在转移居民的任务中发挥了关键作用。岛上有些人被转移到附近的安提瓜和巴布达，也有不少人直接去了英国。

穹状火山

蒙特塞拉特火山的熔岩十分黏稠，无法轻易地从火山通道流出。因此，它们堆成了巨大的穹形块状物。火山喷发经常以火山灰云和火山碎屑流的形态喷射出固态岩浆。

炽热的火山流

蒙特塞拉特火山喷发最危险的特点是火山碎屑流：炽热的火山灰奔流而下，烧掉前进路上的一切东西。留下来试图保卫家园的人们认为他们能猜测出火山碎屑流何时到来，从而化险为夷。但不幸的是，他们并非总是对的。

为什么住在火山附近？

今天，有近5亿人生活在活火山上或者活火山附近。甚至有些大城市就紧邻活火山。例如，墨西哥城靠近波波卡特佩特火山，意大利的那不勒斯靠近维苏威火山。在活火山附近居住是有危险的，但对于生活在这里的居民而言却是利大于弊。

你知道吗？

埃特纳火山是世界上最活跃的火山，它喷发时抛出的熔岩量最多。火山（volcano）这个名字就是由它而来，因为古罗马人相信这里是火神伏尔甘（Vulcan）的家。

在火山的摇篮中

在俄罗斯的远东地区堪察加，180,000人居住在港口城市堪察加彼得罗巴甫洛夫斯克，这座城市周围全是活火山。火山美丽而危险，但是船舶可以由这座港口城市驶入太平洋，人们可以在太平洋上发展渔业，为此居民们甘愿冒险。

进入地狱

矿工在东爪哇伊真火山的火山口里工作，他们采集硫黄，用它们制造火柴和肥料，或者硫化橡胶，硫化的橡胶可以做汽车轮胎。即使火山没有喷发，火山口的水酸性仍然很大，能够腐蚀衣物和金属，甚至损伤肺部！

角马育婴场

Ol Doinyo Lengai在肯尼亚马赛人的语言中意为“神山”。它是一座活火山，活动时会喷发出一种独特的“碳酸岩”灰，使火山周围的土壤十分宜于草的生长。肥沃的草场让角马在一年一度的长途迁徙中止步于此，在此繁育后代。草场也同样吸引了马赛牧民。

酿制美酒的火山

许多葡萄酒生产者说，用在火山土壤上种植的葡萄酿酒是最好的。火山灰会形成重量轻、排水好的土壤，而且富含镁、钙、钠、铁、钾等营养物质。因此，西西里的葡萄种植者甘愿冒险在埃特纳火山的山坡上种植葡萄。

预知火山喷发的前兆

火山研究专家称作火山学家，他们的任务是了解火山，在火山喷发前探知喷发迹象，向附近的居民发出警报。他们搜寻各种能够预示火山喷发的信号，例如，火山底部岩浆房内压力的增加情况，或火山释放出的异常气体。但是，监测火山是一件危险的事情。

你知道吗？

熔岩的化学成分变化可能预示着火山即将喷发。因此，火山学家需要穿着防护服，从新鲜、炽热的熔岩流中取样。

利用无人机

被称为无人机的机器人飞行器逐渐成为火山学家装备库中的重要武器。它们可以飞入没人敢于涉足的火山口。在那里，无人机可以探测到火山要喷发的最清晰迹象。

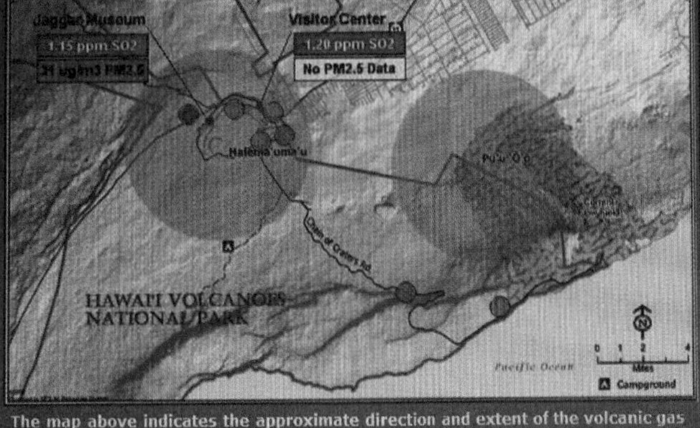

Hawai`i Volcanoes National Park
Current Sulfur Dioxide (SO_2) and Particulate ($PM_{2.5}$) Conditions
Jaggar Museum
1.15 ppm SO2
31 ug/m3 PM2.5
Visitor Center
1.20 ppm SO2
No PM2.5 Data
Halema'uma'u
Pu'u O'o
HAWAI'I VOLCANOES NATIONAL PARK
Pacific Ocean
Miles
Campground
The map above indicates the approximate direction and extent of the volcanic gas plumes (yellow wedges) from Halema'uma'u and Pu'u O'o. Within the plumes are unhealthy concentrations of sulfur dioxide (SO_2) and small particles ($PM_{2.5}$). Color-coded health advisories (color circles or label boxes) are issued when the plumes affect surface areas.
ulfur dioxide gas plumes cross roads near Halema`uma`u and low on hain of Craters Road. Sensitive individuals should limit exposure in these

探寻气体

火山即将喷发时，岩浆沿通道上升，同时释放出二氧化硫和硫化氢等气体。如果火山学家能足够接近火山，他们也许能从火山边上的“烟囱”（火山喷气孔）上，或者是火山口湖里的化学物质中，探测到这些气体的增长情况。

地面位移

火山形状的轻微变化可能说明火山下的岩浆压力正在积聚。计量工具被放置在火山上，借助于GPS卫星，它们能够探测到即便是几毫米的地面形状变化。即使这样微小的变化也可能预示火山即将喷发。

地面震动

在许多火山学家看来，“地震活动性”是预示火山喷发将至的关键线索。“地震活动性”指的是火山表面出现少许晃动和声响。现在，火山专家持续监控着世界上几乎所有主要火山的震动情况，地震仪被放置在离火山不远的地方，或者就在火山的山坡上。

喷发的对比分析

火山学家使用火山喷发指数来衡量火山喷发的强烈程度。火山喷发指数通过诸如火山灰的体积和云团高度等指标来标注喷发的强烈程度——从温和（0）到超级强烈（8）。

海底的大火山

许多火山海拔较高，很难被人们忽视，比如富士山或乞力马扎罗山。但还有许多其他的火山人们几乎看不见，它们在海里喷发。实际上，人们认为在海底有100万座甚至更多的火山，它们喷发出的岩浆超过地球上全部火山喷发产生的岩浆的四分之三。

目击者

1963年11月14日，渔船“伊斯莱维尔二号”上的船员看到冰岛南部海域中有黑烟。起初，他们认为是船只失火产生的黑烟，但事实并非如此，那是一个新岛屿，苏尔特塞岛正在形成。几天之后，这个新岛长度超过了500米，海拔达到了45米。

大西洋中脊在冰岛辛格韦德利峡谷地段高出了海平面

地球上的裂缝

多数海底火山出现在海底大裂缝附近，这些大裂缝是地表的巨大“地壳板块”慢慢拉开之处。在这不为人所见的大洋深处，这些火山创造了世界上最长的山脉——大西洋中脊，它长达40,000千米，从大西洋洋底延伸进入印度洋。

黑烟囱

黄石公园“老忠实”间歇泉是喷发热水的火山管，由于地下火山活动的压力，水在地下加热之后被挤压出来，像这样的火山管被称作间歇泉。令人惊奇的是，海底也有间歇泉，即“海底热泉”。当“海底热泉”喷射出的水与大量的深色矿物质混合在一起，就形成了“黑烟囱”。

热巨人

夏威夷的冒纳凯阿火山高达10,000米，比海拔8844米的珠穆朗玛峰还要高！但冒纳凯阿火山大半部分在海面以下。冒纳凯阿是一座“热点”火山。在这些“热点”，热熔岩的喷泉自地球内部向上，烧熔地壳喷涌而出，这种情况十分罕见。

新生岛屿

当海底火山达到一定高度，喷发物能喷出海面时，海洋中也许一夜之间就会出现新的岛屿。例如，2009年太平洋上的洪阿汤加岛，1973年日本的西之岛，还有1963年冰岛的苏尔特塞岛，这些岛屿都是海底火山喷发后形成的，其中苏尔特塞岛的形成最为壮观。

火山大喷发

在地球的历史上，时不时地发生重大的火山喷发事件，造成世界性的大灾难。2.52亿年前，火山喷发污染了空气，引起大规模的全球变暖，大概有96%的海洋生物和70%的陆地生物灭绝了。如此可怕的事情会再次发生吗？

世界最大的灾难

导致地球上如此多生物灭绝的可怕事件被科学家称为“二叠纪-三叠纪大灭绝”，或简称“大灭绝”。有人认为这是一颗巨大陨石撞击地球造成的。但多数人认为这是由西伯利亚发生的一系列连续的火山大喷发导致的。这些火山喷发使西伯利亚200万平方千米的地区覆盖了熔岩，被称作“西伯利亚地盾”。火山喷发释放出的气体和火山灰在空气中形成了酸雨，并把大量的太阳热量困在大气层中，把整个世界“烤焦了”。

还会发生吗？

将来肯定会有一场灾难性的“超级火山”喷发，就像黄石火山（见下图）一样。尽管没有办法阻止它的发生，但科学家在努力探索监控地下岩浆压力的方法，以便预测喷发是否已经迫在眉睫。多数专家认为，在可以预见的将来不会有超级火山喷发。只是这个可以预见的将来只不过是十年的时间……

人类能逃脱吗？

大约74,000年前，印度尼西亚的多巴火山喷发，它是史上最大的火山喷发之一。喷发将大量的火山灰尘抛入空中。这些火山喷发物飘向全世界，遮蔽了太阳，将世界带入严寒之中。据说只有少数人——大概几千人——熬过了这段可怕的时期。也许正因为这场灾难，让我们能够往上追溯到相同的祖先。

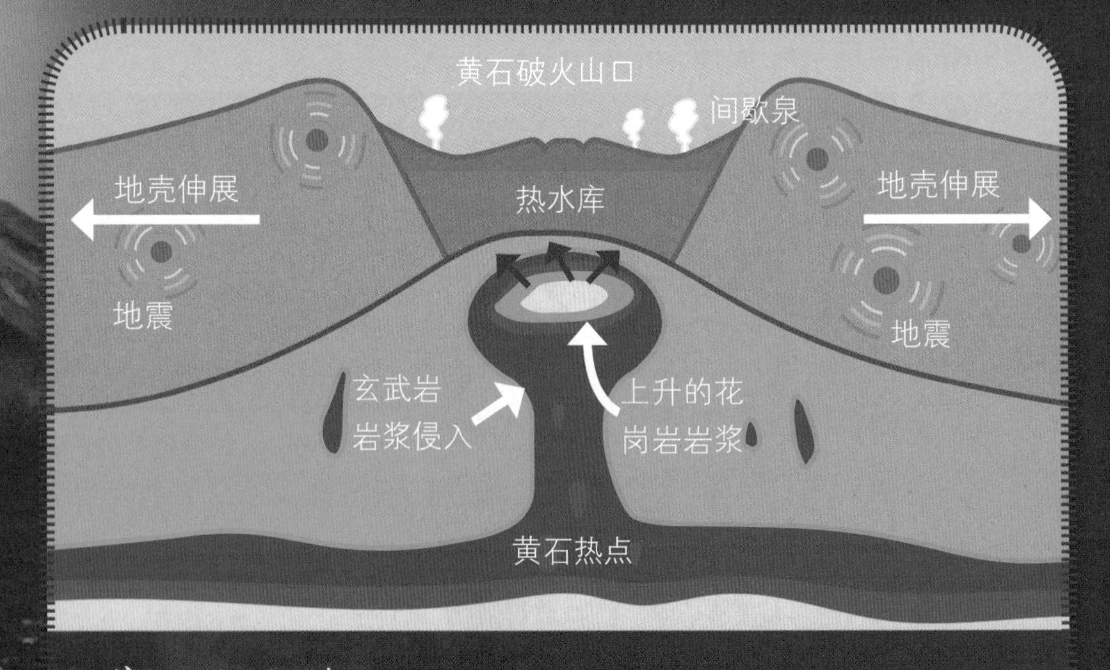

黄石超级火山

事实上，黄石火山是世界上最大的火山之一。火山学家称之为“超级火山”。它的上一次喷发是在64万年前，淹没这一地区的熔岩足以填埋掉美国大峡谷。今天，黄石火山看起来十分温和，但现在已经到了再一次喷发的时候了。科学家认为，如果它真的喷发了，整个美国会被覆盖1米多深的火山灰。

大事件时间轴

公元前1645年—公元前1500年

地中海里的火山岛屿锡拉岛发生了史上最大的火山喷发之一

公元79年

意大利维苏威火山喷发，厚厚的火山灰掩埋了罗马城市庞贝和赫库兰尼姆

6500万年前

现在的印度南部发生了一次大的熔岩喷发，形成了德干地盾，这是一处相当于得克萨斯州面积两倍大的厚厚的熔岩高原

1600年

秘鲁的埃纳普蒂纳发生了美洲历史上最大的火山喷发，摧毁了阿雷基帕城，火山灰云影响了全球气候，并导致俄国发生饥荒

1815年

印度尼西亚坦博拉火山喷发是近代史上最大的火山喷发，抛射出的大量火山灰造成了世界范围的低温夏天

6500万年前

27,000年前

新西兰北岛发生超级火山喷发，形成巨大的火山口，即今陶波湖

1883年

印度尼西亚巽他海峡中的喀拉喀托火山喷发发出的巨响，在几千千米之外都能听到。火山喷发还引发了巨大的海啸

64万年前

一次超级火山喷发（再次）造就了黄石火山地区

1783年

冰岛拉基火山喷发，散发的有毒气体笼罩整个西欧地区

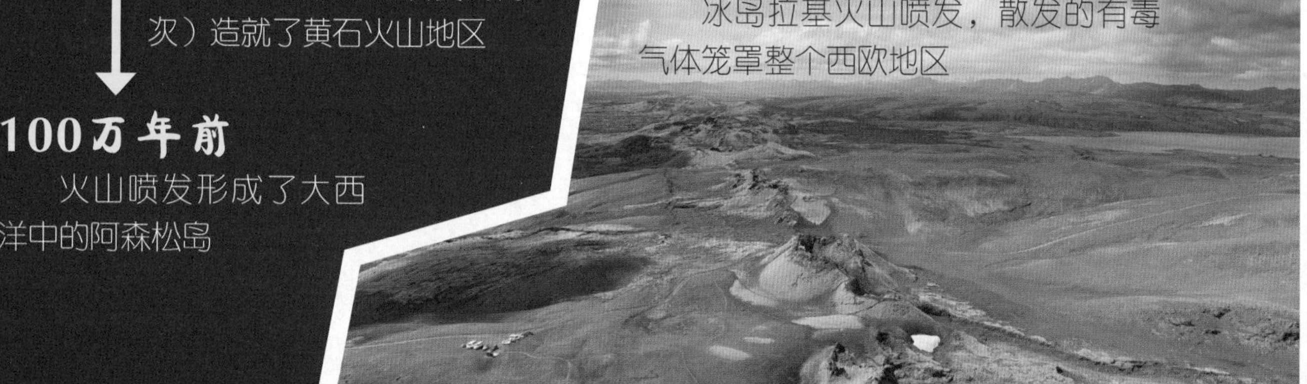

100万年前

火山喷发形成了大西洋中的阿森松岛

1902年

位于加勒比海的法属岛屿马提尼克岛上的培雷火山喷发，它产生的火山碎屑流夺去圣皮埃尔约29,000人的生命

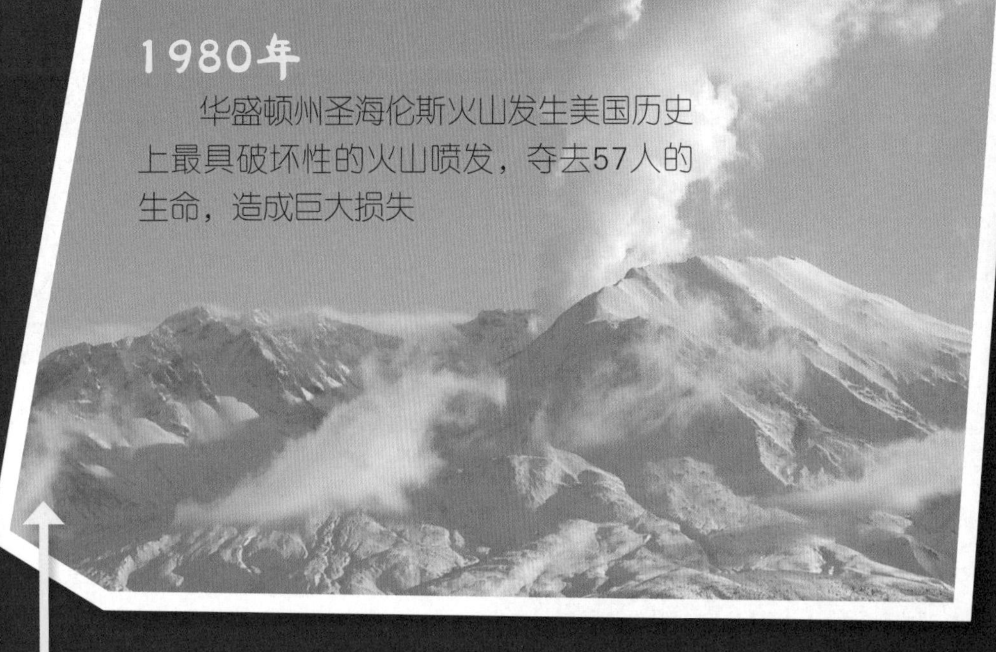

1980年

华盛顿州圣海伦斯火山发生美国历史上最具破坏性的火山喷发，夺去57人的生命，造成巨大损失

1963年

冰岛的苏尔特塞岛在一次海底火山喷发后形成

1985年

哥伦比亚鲁伊斯火山喷发引发的泥石流夺去25,000人的生命

1991年

菲律宾皮纳图博火山喷发，向空中喷射出4立方千米的物质，升起的火山灰柱高达35千米

1982年

墨西哥埃尔奇琼火山喷发

1983年

夏威夷基拉韦亚火山开始喷发

2010年

冰岛埃亚菲亚德拉冰盖下的火山喷发，释放出的火山灰迫使西欧全部客机停飞

爪哇默拉皮火山喷发，迫使35万人逃离家园

1912年

阿拉斯加的诺瓦拉普塔火山喷发是20世纪最大的一次，这次喷发还用尽了附近卡特迈火山的岩浆，导致其坍塌

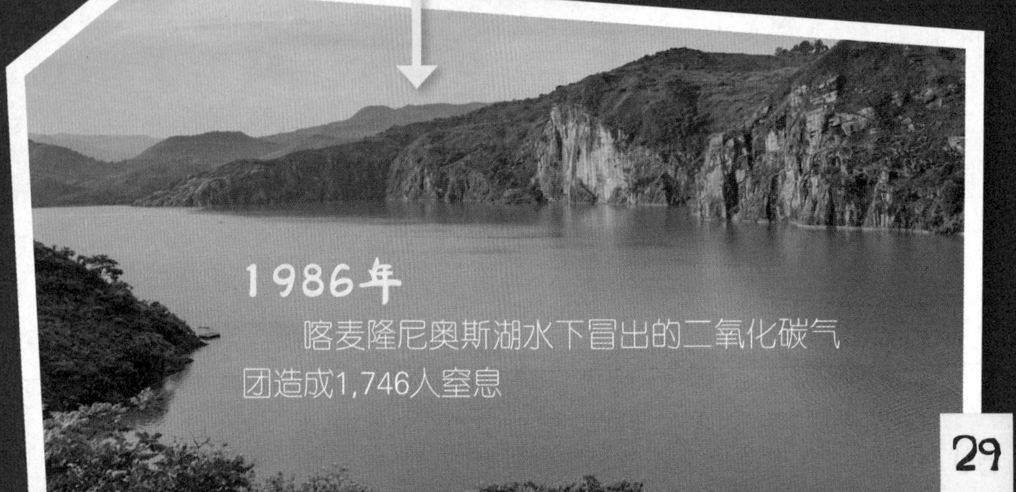

1986年

喀麦隆尼奥斯湖水下冒出的二氧化碳气团造成1,746人窒息

高温和喷发

关于火山的惊人事实

大怪兽

尽管地球上有一些大火山，但太阳系中最大的火山在火星上，名为奥林匹斯山。它硕大无比，约有600千米宽，21千米高。

高峰

离地心最远的点不是珠穆朗玛峰的顶峰，而是厄瓜多尔的钦博拉索山的山顶。这是因为地球旋转得很快，因此它的中部是凸出的。这就使得地球的赤道部分更厚一些，而钦博拉索山就非常靠近赤道。

木卫一

木星的卫星木卫一（见上图，在它的母星前面）是已知火山活动最活跃的地方。这颗卫星的表面就是一个充满了火山活动的锅形陷坑。实际上，木卫一的表面由于火山喷发而时刻在变化。

浮石

浮石是一种轻且多孔的岩石。当玻璃质的泡沫熔岩冷却、固化时，浮石就形成了。这种石头真的可以漂浮！

正在发生

此时此刻，世界上至少有20座火山正在喷发之中。每年有50座到70座的火山发生喷发，过去的十年里，大约160座火山处于活跃状态。科学家估计，在过去的一万年中已有1,300座火山喷发。

火山孵化

在印度尼西亚的苏拉威西岛上，马累家雉不是伏在它们的卵上给卵保温，而是聪明地利用火山的热量帮助它们孵卵！它们把卵埋进靠近火山的土或沙中，雏雉孵化后用爪子扒开土或沙，走出地面。

炽热的熔岩

有时熔岩的流动速度也许不是很快，但即便如此，你也不会想挡在它面前。它的温度可达1,250摄氏度，几乎能够烧毁一切事物！

火山的世界

世界上的很多东西要归功于火山。超过80%的地壳是火山造成的，尽管大部分地壳现在覆盖着一层沉积物，但是大部分沉积物都是破碎了的火山岩碎片。海底和一些山岭是由无数次的火山喷发形成的。火山释放的气体还参与了地球大气层的构成。

火山喷发之最

历史上最大的爆炸性喷发
巴西，巴拉那地盾，瓜拉普阿瓦-塔玛拉那-萨鲁萨斯
1.32亿年前
体积：8,600立方千米——足以覆盖纽约州60多米深

2500万年之内最大的喷发
印度尼西亚巽他岛弧，多巴湖
74,000年前
体积：2,800立方千米——足以覆盖纽约州20多米深

近5万年最大的喷发
新西兰，陶波
27,000年前
体积：1,170立方千米——足以覆盖纽约州8米多深

最大的熔岩喷发
印度德干地盾，默哈伯莱什沃尔-拉贾蒙德里地盾
6,500万年前
体积：9,300立方千米——足以覆盖纽约州70多米深

最致命的火山喷发
1815年，印度尼西亚坦博拉火山
死亡人数：92,000人

上个世纪最致命的火山喷发
1902年，马提尼克岛培雷火山
死亡人数：29,000人

近50年内最致命的火山喷发之一
1991年，菲律宾皮纳图博火山
死亡人数：300人

最致命的火山效应
1600年，秘鲁，埃纳普蒂纳
死亡人数：巨大的火山灰云造成全世界的极冷气候，200万俄国人死于饥荒

作者简介

约翰·法恩登，英国伦敦城市行业协会的皇家文学基金会研究员。他为成人与儿童写作了大量图书，涉及科学、技术、自然等领域，包括畅销世界的《你觉得自己聪明吗？》和《放我出去》。他六次入围英国皇家学会青少年科学图书奖，入围书目有《地球原理》《何因何果》和《人体项目》（2016）等。

WHEN DISASTER STRIKES
EXTREME HURRICANES AND TORNADOES

Written by John Farndon

外语教学与研究出版社
FOREIGN LANGUAGE TEACHING AND RESEARCH PRESS
北京 BEIJING

Imagine facing a wall of water 9 metres high or winds of 300 km/h. When an extreme storm hits, one thing is certain: the damage and destruction will be extreme. The unleashed power can flatten homes, drown low-lying land in torrential rain, and throw cars about like toys.

Discover the worst that hurricanes and tornadoes can do, how and where they happen, and about the drone and satellite technology used for tracking them – and about the brave storm chasers who race towards danger...

HURRICANE
SEASON

CONTENTS

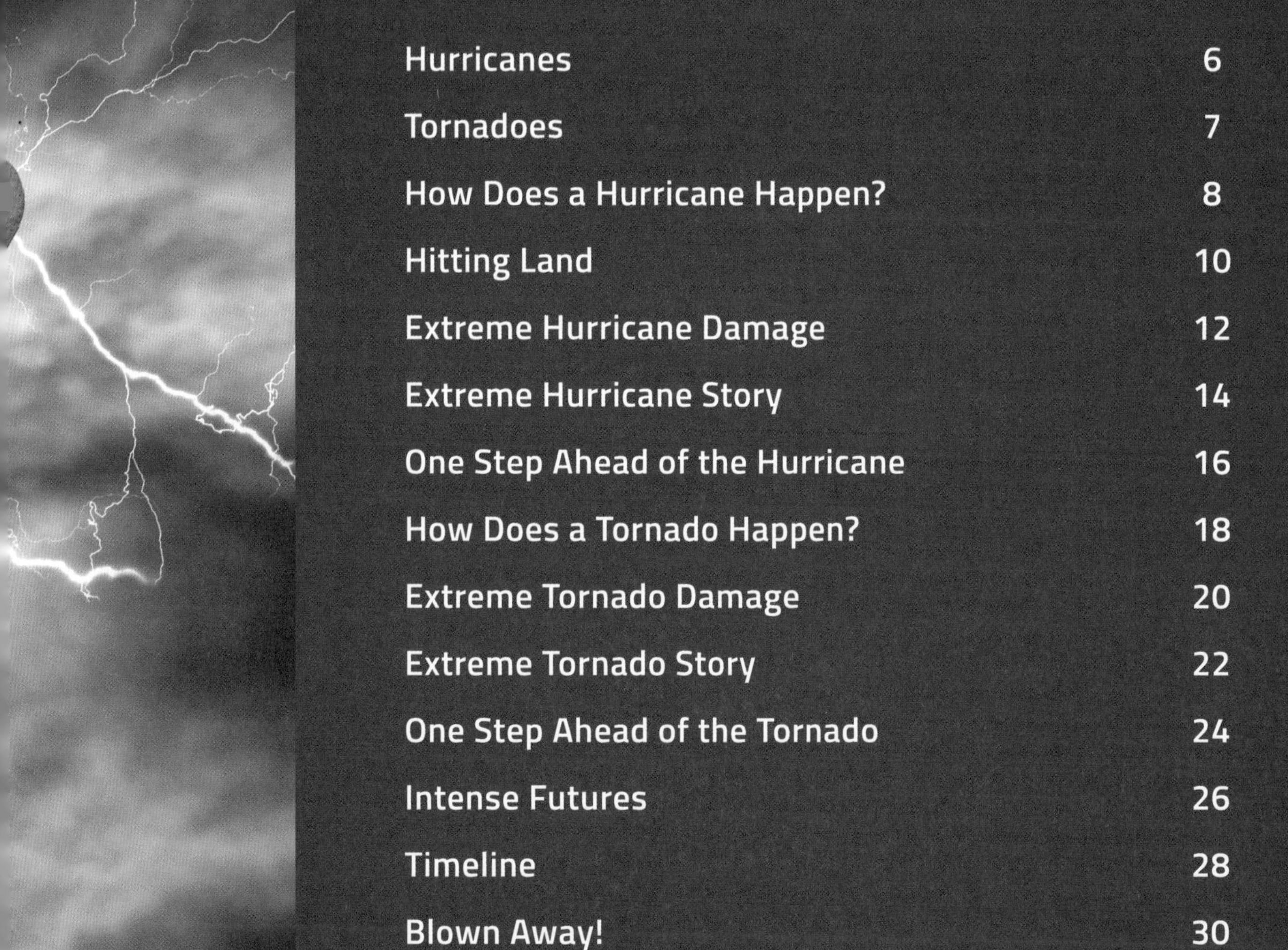

HURRICANES

A hurricane is a storm so vast you can only see it all from space. From high above, it looks like a huge whirling cream cake. The white is rings of huge, powerful thunderclouds that bring torrential rain. The whirls are created by powerful winds that blow in a spiral round the storm's centre or 'eye'.

TROPICAL CYCLONES

'Hurricane' is the name given to a spiralling tropical storm. Similar storms may also be called typhoons in the Pacific. Scientists call them both 'tropical cyclones'.

TORNADOES

Tornadoes, also known as twisters and whirlwinds, are spinning, roaring columns of air just a few hundred metres across. They come down from giant thunderclouds and rip across the landscape, destroying buildings, uprooting trees and hurling cars in the air. Winds spiral around the outside at ferocious speeds – while pressure in the centre is so low it can suck objects up like a giant vacuum cleaner.

HOW DOES A HURRICANE HAPPEN?

Hurricanes are stirred into life in late summer by tropical sun beating down on the ocean and steaming off water to build giant thunderclouds. High above, strong winds blow from the east. The winds skim the cloud tops and swirl them together in one big spiral storm. The storm wheels westwards, gaining power as it gathers in more clouds.

UNBELIEVABLE!
According to NASA, a hurricane can expend as much energy during its life cycle as 10,000 nuclear bombs!

CROSS-SECTION OF A HURRICANE

Rotating cloud mass

In the very centre of the eye, air sinks again, creating a brief moment of clear skies and calm weather

EYE

Winds escape through the eye

Winds roar round anticlockwise at sea level

ANATOMY OF A HURRICANE

In a hurricane, rain lashes down from rings of thunderstorms known as rainbands, while howling winds at sea level drive the storm anticlockwise. Right in the centre there is a clear tunnel up through the clouds, called the 'eye', where winds spiral up the cloud walls and out.

WHIRL OF POWER

A big hurricane releases as much energy in a day as all the world's power stations in a year. The power comes from all the thunderclouds clustered together in a spiral of cloud rings hundreds of kilometres across.

HURRICANE FORCE

Hurricanes create superstrong winds. To be classed as a hurricane, a storm must have winds of at least 118 km/h. But in a powerful hurricane, winds can get much stronger. In Hurricane Camille in 1969, winds reached 305 km/h!

STORM FROM AFRICA

In the Atlantic, hurricanes typically begin off Africa, near the Cape Verde islands. As they develop, they move west over the ocean at about 24 km/h. In under two weeks, they hit the Caribbean and swing northwards. By this time they are at the height of their power.

A HURRICANE'S BIRTHPLACE

Hurricanes often begin in the North Atlantic Ocean or the eastern North Pacific. They sweep westwards as they develop, then swing away from the Equator before finally petering out.

HITTING LAND

It's once they hit land that hurricanes really do their damage. Long before the storm arrives, huge waves stirred up by the wind start to smash against the coast. Then watchers on the shore see ominous dark clouds heading their way, and feel the wind picking up. They're in for a rough time!

UNBELIEVABLE!

When Hurricane Katrina hit Waveland, Mississippi, USA on 29 August 2005, the storm surge was more than 7.9 m. That's not just a wave – it means the entire sea rose higher than a house!

HURRICANE IN HAITI

When Hurricane Matthew hit Haiti (main picture) in 2007, the winds were strong enough to bend even the strongest trees, and blow roofs off all but the stoutest buildings. They are also strong enough to blow away a big car. Torrential rains brought instant flooding.

TRICKED!

After hours of battering, the onslaught may often seem to die down. The rain stops. The wind drops. The sun may come out. But this is a brief respite as the eye of the storm passes over. Within an hour, the eye has moved on and the rain and wind come storming back.

STORM SURGE

Low air pressure in the hurricane's eye lifts the ocean surface up in a dome. Winds pile up water even higher. This is called a 'storm surge'. As the hurricane moves landwards, it drives the surge with it, creating a massive high tide that can swamp coastal areas and sweep far inland.

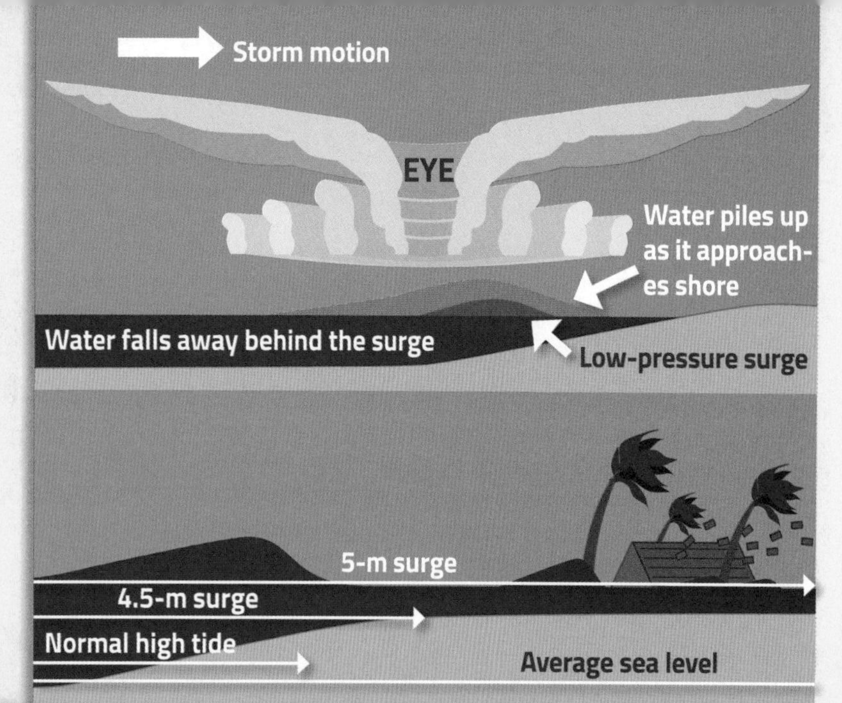

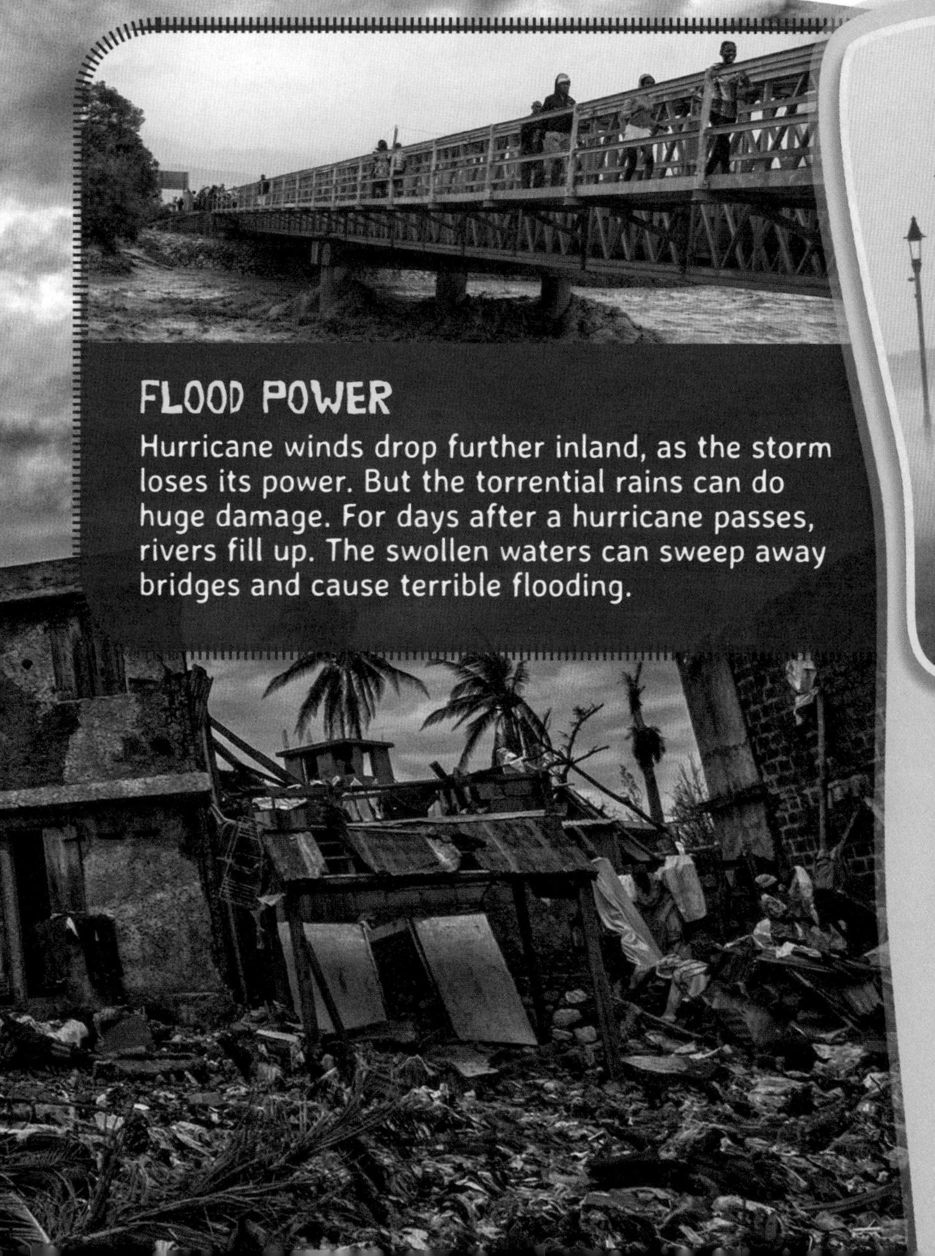

FLOOD POWER

Hurricane winds drop further inland, as the storm loses its power. But the torrential rains can do huge damage. For days after a hurricane passes, rivers fill up. The swollen waters can sweep away bridges and cause terrible flooding.

WAVE POWER

Coasts exposed to hurricanes take a real battering from giant waves. When Hurricane Ivan hit the Atlantic coast of the USA on 15 September 2004, it threw up waves over 27 m high. Just imagine a wall of water as high as a 10-storey building coming at you!

EXTREME HURRICANE DAMAGE

Some hurricanes cause little more than inconvenience. But the most powerful can be hugely destructive, especially if they hit areas where a lot of people live. The ferocious winds can destroy buildings and power lines, while the floods from the heavy rain can wash away bridges, cut off roads and railways, and trap people.

UNBELIEVABLE!

A slower-moving, weaker hurricane can often do more damage than a faster-moving, more powerful one, because it stays long enough to drop heavy rain and cause flooding.

SANDY MONSTER

Hurricane Sandy of October 2012 (main picture) was one of the biggest hurricanes, measuring 1,450 km across. It caused at least $75 billion damage when it hit land everywhere from Jamaica to the USA — second only to Hurricane Katrina.

GALVESTON CALAMITY

On 8 September 1900, the city of Galveston (right) in Texas, USA, then the 'New York of the South', was utterly destroyed when a hurricane ripped through it. The 225-km/h winds flattened many buildings, while a 4.5-m storm surge washed them away. More than 6,000 people died and 3,600 buildings were totally destroyed.

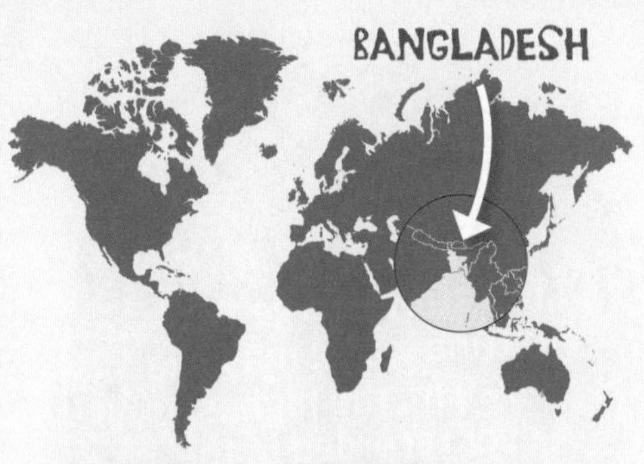

SHIP TO SHORE

There were few better shows of the power of Hurricane Sandy than the appearance of a 52-m tanker on Front Street on New York's Staten Island. The ship had been anchored 1.6 km out in the bay, but was hurled onshore by the storm.

FLOOD TERROR

More than half of Bangladesh lies less than 6 m above sea level. Worse still, the shape of the Bay of Bengal means storm surges are funnelled towards the country's low-lying coast. So cyclones have repeatedly brought floods to Bangladesh, with devastating effects.

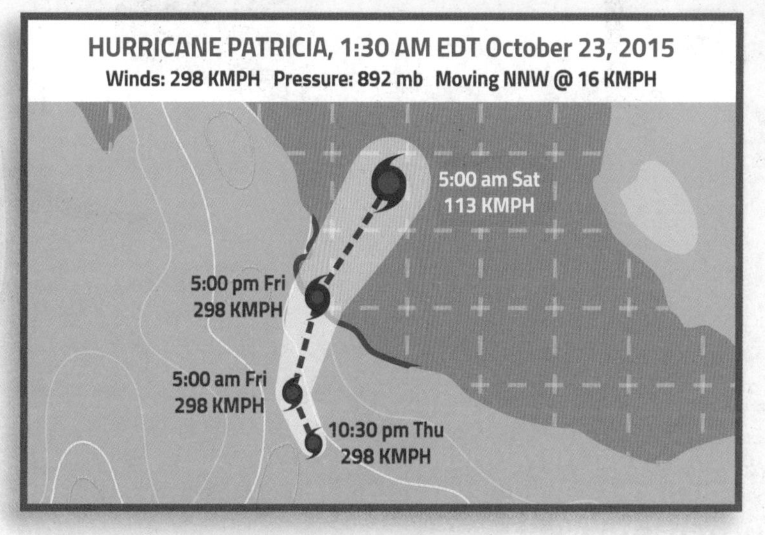

POWERHOUSE PATRICIA

October 2015's Hurricane Patricia was the second most intense hurricane ever – with the second lowest pressures ever recorded, at its centre. It hit land in Mexico with ferocious winds. Fortunately, Patricia ran straight into mountains and rural areas, so did much less damage than if it had hit a city.

EXTREME HURRICANE STORY

No one who lived in New Orleans, Louisiana, USA, in August 2005 will ever forget Hurricane Katrina. It was one of the deadliest, and certainly the costliest, natural disasters ever to hit the USA. More than 1,800 people lost their lives, and estimates put the cost of the damage at well over $160 billion.

UNBELIEVABLE!
The storm surge from Katrina was a massive 5–9 m, the highest ever recorded in the USA.

BURST BANKS

Low-lying New Orleans relies on banks called 'levees' to keep water within its waterways. But during Katrina, the storm surge and heavy rain combined to fill the waterways to bursting. Soon water spilled over the levees, then breached them altogether in 50 different places, flooding the city.

DEMOLITION

These pictures show a beach house in New Orleans before and after Katrina hit. Altogether, over a million homes were damaged, 134,000 in New Orleans alone – that's nearly three-quarters of all homes in the city. The main cause of the damage was flooding.

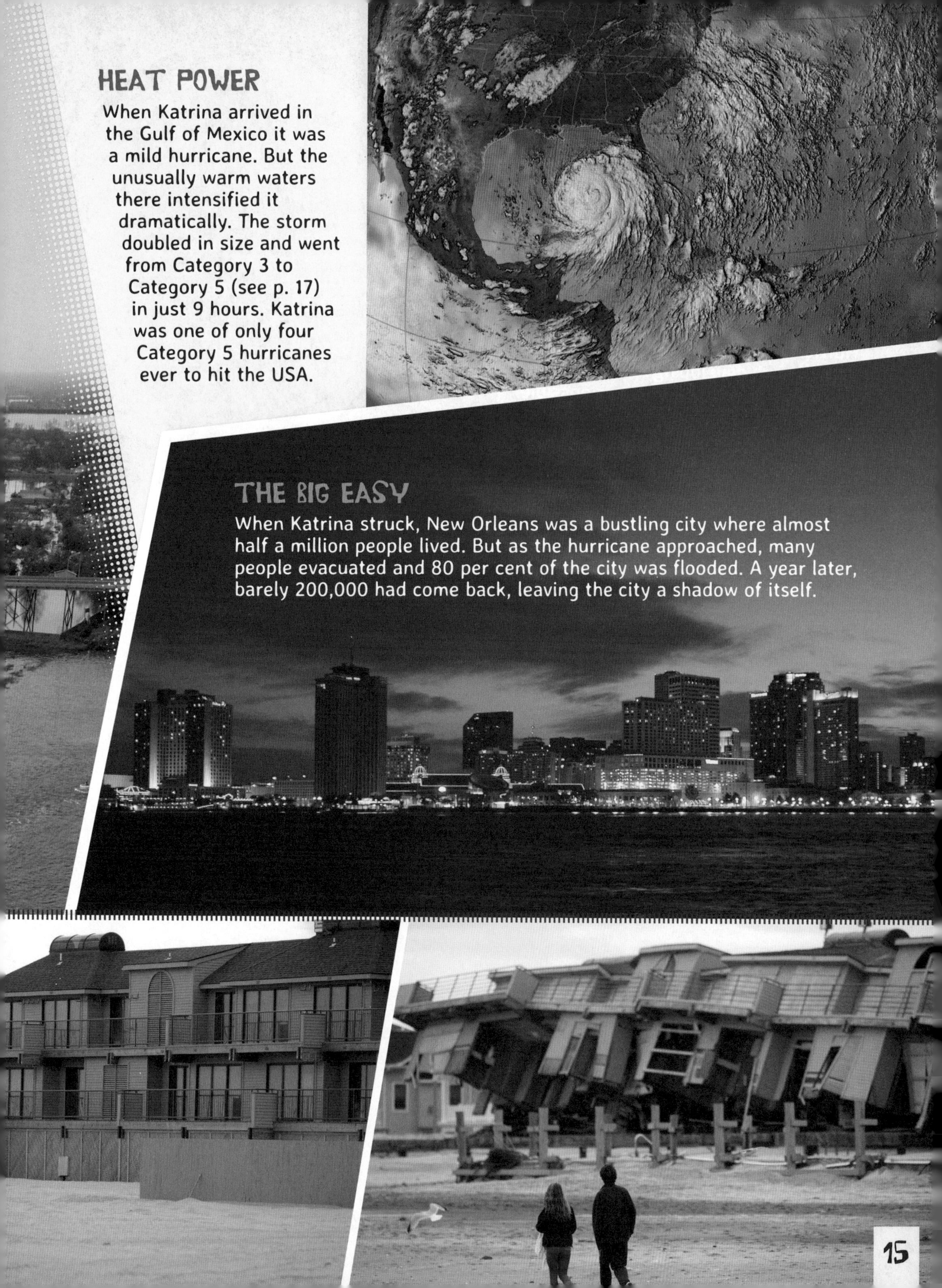

HEAT POWER

When Katrina arrived in the Gulf of Mexico it was a mild hurricane. But the unusually warm waters there intensified it dramatically. The storm doubled in size and went from Category 3 to Category 5 (see p. 17) in just 9 hours. Katrina was one of only four Category 5 hurricanes ever to hit the USA.

THE BIG EASY

When Katrina struck, New Orleans was a bustling city where almost half a million people lived. But as the hurricane approached, many people evacuated and 80 per cent of the city was flooded. A year later, barely 200,000 had come back, leaving the city a shadow of itself.

ONE STEP AHEAD OF THE HURRICANE

Every year between June and November, about 10 hurricanes sweep across the Atlantic – and one or two strike the coast of the USA. This is the hurricane season and the National Hurricane Centre in Florida , USA, is on high alert to track developing hurricanes and give people as much warning as possible.

UNBELIEVABLE!

Hurricane Hunters (right) sometimes fly right through the hurricane's eye wall, the most intense part of the storm! The planes are lashed by driving rain and hail, battered by winds of 240 km/h or more, and tossed by violent updrafts and downdrafts.

SPACE EYE VIEW

Scientists can follow the path of hurricanes easily from satellites high above. By putting several hours of satellite pictures together, they can see how it is developing. But hurricanes can change path very quickly.This is Hurricane Matthew threatening Florida in October 2016.

THE HURRICANE HUNTERS

Drones, or robot aircraft, can be guided into a hurricane by remote control. They can be sent to monitor conditions in the most dangerous parts of a hurricane without putting aircrew's lives at risk. Drones can fly for 30 hours at twice the height of a passenger plane.

Doppler radar detects things by bouncing microwaves off them. With a special kind of Doppler radar, hurricane watchers can build up a detailed map of where and how much rain is falling. It can even give an indication of wind speed. This radar station is located in Dodge City, Kansas, USA.

The strength of a hurricane is graded in categories from 1 to 5 on the Saffir-Simpson hurricane wind scale.

Category	Wind Speed	Damage
1	119–153 km/h	Light damage: mobile homes shifted; signs blown over; branches broken
2	154–177 km/h	Moderate damage: mobile homes turned over; roofs lifted
3	178–208 km/h	Extensive damage: small buildings wrecked; trees uprooted
4	209–251 km/h	Extreme damage: most trees blown down; widespread structural damage to all buildings
5	Above 252 km/h	Catastrophic damage: most buildings destroyed; forests, roads and pipelines wrecked

HOW DOES A TORNADO HAPPEN?

The most dangerous place in the world for tornadoes is Tornado Alley, which stretches across North America's Great Plains. Here, warm air from the Gulf of Mexico collides with cold winds blowing from the Rocky Mountains. This creates huge thunderstorms called supercells, which are like mighty tornado makers.

UNBELIEVABLE!

Typically the winds whipping round a tornado are about 160 km/h. But sometimes they can reach as much as 480 km/h and have the power to pick up a log and hurl it through a brick wall!

BANGS AND FLASHES

The approach of a tornado is terrifying. The spinning winds roar like a very loud freight train, while thunderclaps boom and lightning flashes from the cloud above. Fortunately, they only tend to last 10 minutes or so before petering out.

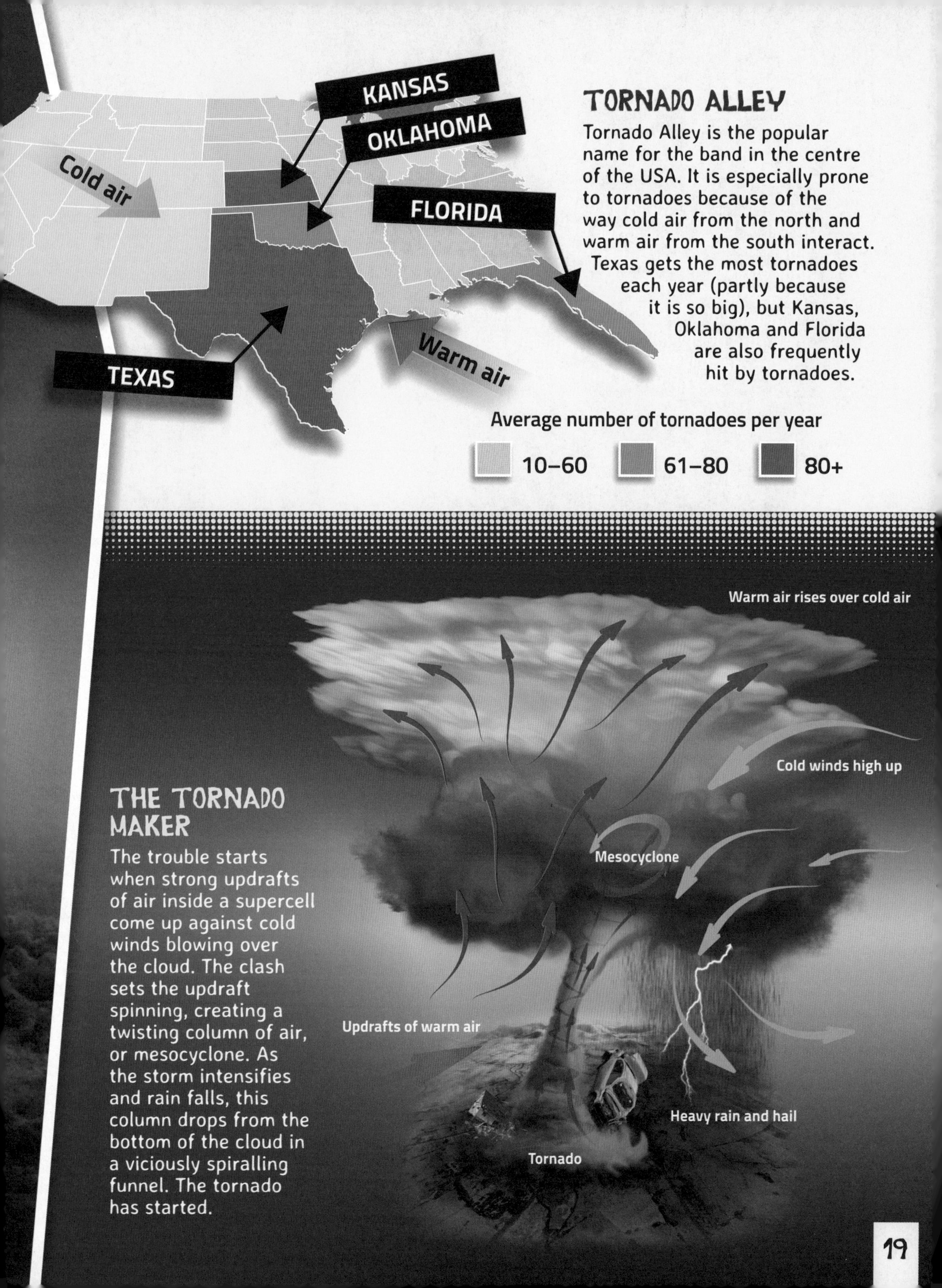

TORNADO ALLEY

Tornado Alley is the popular name for the band in the centre of the USA. It is especially prone to tornadoes because of the way cold air from the north and warm air from the south interact. Texas gets the most tornadoes each year (partly because it is so big), but Kansas, Oklahoma and Florida are also frequently hit by tornadoes.

THE TORNADO MAKER

The trouble starts when strong updrafts of air inside a supercell come up against cold winds blowing over the cloud. The clash sets the updraft spinning, creating a twisting column of air, or mesocyclone. As the storm intensifies and rain falls, this column drops from the bottom of the cloud in a viciously spiralling funnel. The tornado has started.

EXTREME TORNADO DAMAGE

Often, tornadoes whirl by doing little damage, apart from shaking a tree or blowing a loose door off its hinges. But sometimes they cause catastrophic damage, blasting buildings flat, hurling cars through the air, and killing and injuring people and animals. The severity of a tornado is measured from 0 to 5 on the Enhanced Fujita scale.

UNBELIEVABLE!

The deadliest tornado in US history occurred on 18 March 1925. It killed 695 people and travelled more than 480 km through Missouri, Illinois and Indiana. The storm was rated an F5 on the Fujita scale.

HOUSE DOWN

This house in Joplin, Missouri, blown apart in May 2011 by an EF5 tornado, reveals the destructive power of tornadoes. Curiously, books are left neatly stacked on the shelves, showing just how localised the effects can be.

EXTREME STORY: RAINING FROGS

Amazingly, rainstorms can often bring showers of frogs as well as water! The theory is that they are picked up by tornadoes skimming over ponds. Tornadoes that form over water, known as waterspouts, can even pick up fish!

TORNADO SCALE

The damage done by a tornado is rated from 0 to 5 on the Enhanced Fujita scale.

Scale	Wind Speed	Damage
EF0	105–137 km/h	Minor damage: roof tiles disturbed; branches of trees broken
EF1	138–177 km/h	Moderate damage: roof tiles torn off; mobile homes overturned; cars blown off roads
EF2	178–217 km/h	Considerable damage: roofs blown off; mobile homes destroyed; railway trucks overturned; large trees uprooted
EF3	218–266 km/h	Severe damage: houses blown down; cars hurled through the air; heavy objects flung like missiles
EF4	267–322 km/h	Extreme damage: houses, even with strong foundations, destroyed; cars turned into flying missiles; trees stripped of bark
EF5	Above 322 km/h	Total destruction: catastrophic, widespread damage

EXTREME TORNADO STORY

On the afternoon of 22 May 2011, the city of Joplin, Missouri, was hit by one of the worst tornado disasters in American history. Warning sirens went off 24 minutes before the tornado struck, but many people took no notice – after all, there had been many false alarms. But this one was for real.

UNBELIEVABLE!

The strike on Joplin was so sudden and so devastating that people were left in total shock. One rescue worker was approached by a dazed-looking man, who asked, "When are they going to turn my power back on?" But his house had been completely destroyed.

FLATTENED

The damage was done by winds that ripped through the town at speeds of 400 km/h, meaning that the tornado was likely to have been an EF5 category. In the flattened hospital, the remains of a large truck were found — it had been carried 115 m and wrapped round a tree stripped of its bark!

BEFORE AND AFTER

Before the tornado hit, Joplin was a quiet town of 50,000 people, with many solid brick buildings that had stood for over a century. But minutes later, up to 7,000 buildings had been reduced entirely to rubble. It was as if an atomic bomb had fallen.

STORM TRACK

The tornado sliced a 1600-m-wide path of devastation through Joplin. Zack Rosenburg, who helped with rebuilding, said, "The track of that tornado was crystal clear. One side of the street would be completely devastated and the other side of the street had no damage at all."

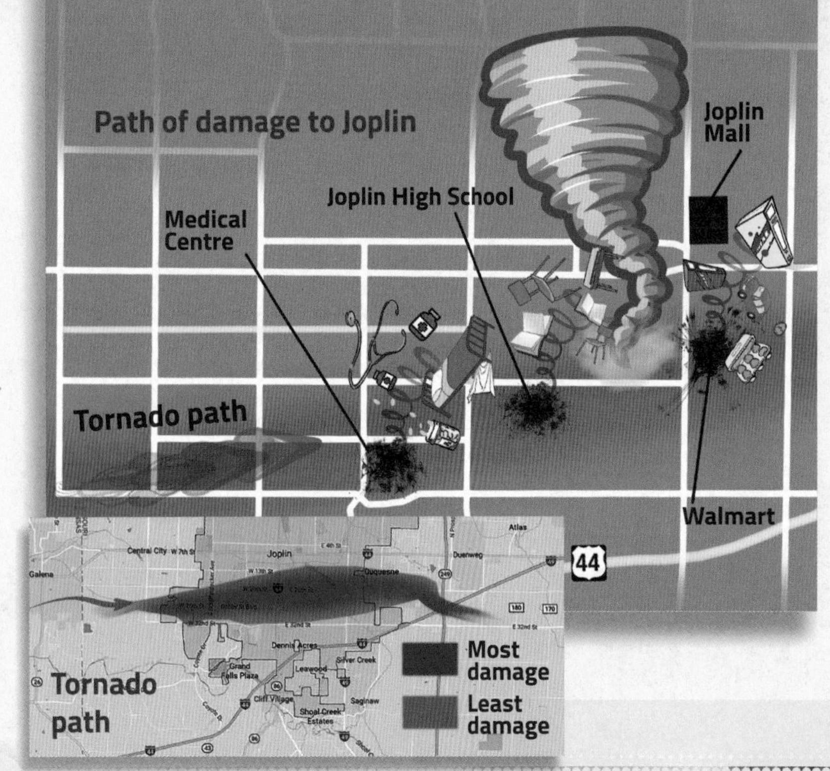

EXTREMES

The Joplin tornado was the seventh deadliest in US history, killing 161 people. The damage done in just a few minutes was so devastating that most experts accorded it an EF5 rating. Trucks and concrete blocks were hurled hundreds of metres, indicating wind speeds of over 320 km/h.

REBUILDING

Don Attebury, aged 89 years, was one of many Joplin residents whose home was destroyed (above). But afterwards the city embarked on a remarkable rebuilding programme, and by 2016 it was hard to see that the tornado had ever happened. Don Attebury's house itself looked as good as new.

ONE STEP AHEAD OF THE TORNADO

In the last half-century, our understanding of tornadoes has increased dramatically, thanks in part to brave storm chasers. Some storm chasers are photographers hoping to capture amazing shots. Others, such as the scientists of the VORTEX2 and ROTATE2012 projects, are hunting for data that will indicate what conditions trigger tornadoes.

EYEWITNESS

"If I close my eyes I can still hear the sound. It was something I'll never forget – like a high-pitched whistling noise and a low rumble at the same time."
Storm chaser Steve Johnson, on coming very close to a tornado.

CAN BIRDS HEAR TORNADOES COMING?

In 2014, scientists noticed that five little golden-winged warblers they were tracking suddenly fled from their nest sites. A day later, a tornado struck. After a few days, with the tornado past, the birds returned safely to their nests. Scientists believe their ultra-sensitive hearing picked up the sound of the tornado in the distance.

TORNADO WARNING

In the 1950s, sirens were set up in many places to warn of air raids in times of danger. Now they are used to warn people a tornado is on its way. Mind you, they don't give you much warning: 13 minutes on average. But that could be just enough for you to get to a place of safety!

THE SOUND OF A TORNADO

Doppler radars have become crucial in tracking tornados. Like conventional radars, they detect things by sending out radio waves and seeing how they bounce back. But Doppler radars can detect which way a storm is moving, as well as its position, from changes in the frequency of waves bounced off water droplets in the storm cloud.

TORNADO CHASERS

American filmmaker Sean Casey became famous for his exploits in the TV series *Storm Chaser*. To get close enough to the action, he built two armour-plated 'Tornado Intercept Vehicles' (TIVs), with a camera turret, weather sensors, bulletproof windows, and drop-down metal skirts to protect the underside of the vehicle against tornado missiles.

INTENSE FUTURES

This century has seen a number of extreme storms, from Hurricane Katrina to Cyclone Nargis, which killed 138,000 people in Myanmar in 2008. But many scientists believe we are in for even more severe storms. Climate change – the changes in the world's climate triggered by polluting gases such as carbon dioxide – may give the atmosphere extra energy and disrupt weather patterns.

RISING COSTS

Eight of the 10 costliest hurricanes on record in the USA have occurred since 2004. Hurricanes Katrina (2005) cost $160 billion and Sandy (2012) $75 billion. That's partly because more people are now living near the sea, but storms also seem to be becoming more ferocious.

TORNADO CLUSTERS

Over the last decade or so, scientists have noticed a change in the pattern of tornadoes. While the number has stayed much the same, tornadoes now seem to be coming in devastating clusters, with long gaps in between. Just such a cluster brought the Joplin tornado of 2011. Some scientists think climate change may be to blame.

STORMY WEATHER

Hurricanes are getting more frequent. Between 1966 and 2009, there were on average six full-blown hurricanes each year. Since 2009, there have been eight hurricanes a year. This is because the ocean has become much warmer. Scientists are not sure why the ocean is warmer, but many believe it is due to climate change.

DANGEROUS BABY

'El Niño' is the Spanish for 'baby boy'. It is also the name of a powerful climate event triggered by changes in ocean currents in the Pacific. Every about two to seven years, warm water piles up in the east of the ocean against South America. The result is more extreme weather around the world. Some scientists think climate change may make El Niño effects even more extreme, bringing more intense storms.

TIMELINE

1054
The earliest recorded tornado in Europe struck Rosdalla, near Kilbeggan, Ireland

1609
A hurricane wrecked the settlers' ship *Sea Venture* (left) on Bermuda, inspiring Shakespeare's famous play *The Tempest*

1281
The Hakata Bay Typhoon in Japan wiped out the entire Mongol fleet, earning it the name Kamikaze or 'divine wind'

1635
The Great Colonial Hurricane was the first known hurricane in New England, USA

1780
The Great Hurricane, one of the deadliest Atlantic hurricanes ever, killed over 20,000 people in the Caribbean

1815
The Great September Gale hit New England

1054

1559
A hurricane destroyed the first colony in Florida

1494
Columbus made the first European record of a hurricane. He first sailed to the Americas in the *Santa Maria* (replica shown)

1839
The Coringa Cyclone hit Southeastern India with devastating effect, killing nearly 300,000 people

1737
The Hooghly River Cyclone, which hit India and Bangladesh, was one of the deadliest storms ever, killing over 350,000 people and destroying 20,000 ships

1667
The worst ever hurricane in Virginia, USA, destroyed 10,000 houses

1881
The Haiphong Typhoon travelled through the Beibu Gulf, in the South China Sea, and killed 300,000 with its storm surge

1974
The Super Outbreak of tornadoes in the USA unleashed 148 tornadoes in just 24 hours

2011
Former US President Barack Obama meets the people of Joplin, devastated by a deadly tornado

1925
The Tri-State Tornado, which hit Missouri, Illinois and Indiana, was one of the biggest and most ferocious ever

1992
Hurricane Andrew was the third most expensive hurricane in US history, causing $26.5 billion damage

1923
Tokyo was destroyed in the fire whipped up by a typhoon that followed an earthquake

2005
Hurricane Katrina was the most expensive hurricane in the USA ever, devastating the city of New Orleans

1988
Hurricane Gilbert was one of the most intense storms ever, causing terrible damage in Central America

1989
The world's deadliest tornado hit Daulatpur-Saturia in Bangladesh, killing 1,300 people

1970
The Great Bhola Cyclone killed 500,000 people in Bangladesh, making it one of the worst natural disasters

1975
Super Typhoon Nina burst major dams in China, driving 11 million people from their homes with the flood

1900
The Galveston Hurricane in Texas was one of the worst to hit America, swamping the city with its storm surge

2010
Hurricane Sandy wreaked havoc along the southeastern coast of the USA up to New York City, seen here

BLOWN AWAY!

Amazing facts about hurricanes and tornadoes

PICK-UPS

There is a story that in Oklahoma, USA, a small herd of cattle were sucked up by a tornado then set down unharmed some distance away. It probably isn't true. But in April 2011, a boy was said to have been plucked from his bed and dumped a few hundred metres from home. Such escapes are rare.

THE DADDY OF THEM ALL

The largest and most powerful hurricanes on Earth may be 1,600 km across, and have winds of 320 km/h. That's pretty awesome. But on Jupiter there is a storm twice as wide as the entire Earth, with winds of more than 640 km/h. What's more, it's been going on at least 150 years. It's called the Great Red Spot.

NAME YOUR STORM!

Every hurricane, cyclone and typhoon is given its own name to help forecasters as they track it. Hurricanes were first given names by an Australian weather forecaster named Clement Wragge in the early 1900s. The first hurricane of the year is given a name beginning with the letter A. When a hurricane is especially devastating, its name is permanently retired and another name replaces it.

BUSTED!

Some people will tell you that you should open windows to let a tornado through safely. This is a myth. Opening the wrong windows can allow air to rush in and blow the house apart from the inside.

PROJECT STORMFURY

Between 1962 and 1983, the US government flew planes over hurricanes in Project Stormfury. The idea was to drop silver iodide dust into the clouds to act as seeds for raindrops to grow and cool the hurricane. The idea didn't work.

EVIL SPIRITS

Tornado comes from the Spanish word *tornada*, meaning thunderstorm. The word hurricane comes from a Taino Native American word, *hurucane*. It means the evil spirit of the wind.

WORST CYCLONES

DEADLIEST

The Great Bhola Cyclone (Bangladesh)
12 November 1970
Cost in lives: up to 500,000

MOST EXPENSIVE HURRICANE

Hurricane Katrina (USA)
29-31 August 2005
Cost in lives: 1,836
Financial cost: $160 billion

STRONGEST WINDS

Cyclone Olivia (Australia)
10 April 1996
Wind speed: 408 km/h

WETTEST

Cyclone Hyacinthe (Indian Ocean)
14-28 January 1980
Rainfall: 6,083 mm

WORST TORNADOES

WORLD'S DEADLIEST

Daulatpur-Saturia (Bangladesh)
26 April 1989
Cost in lives: 1,300

AMERICA'S DEADLIEST

Tri-State Tornado (USA)
18 March 1925
Cost in lives: 695

MOST EXPENSIVE

2011 Super Outbreak (USA)
26-28 April 2011
Cost in lives: 348
Financial cost: $11 billion

BIGGEST RECORDED

El Reno (USA)
31 May 2013
Distance across: 4.2 km

INDEX

THE AUTHOR

John Farndon is Royal Literary Fellow at City&Guilds in London, UK, and the author of a huge number of books for adults and children on science, technology and nature, including such international best-sellers as *Do You Think You're Clever?* and *Do Not Open.* He has been shortlisted six times for the Royal Society's Young People's Book Prize for a science book, with titles such as *How the Earth Works, What Happens When...?* and *Project Body* (2016).

Picture Credits (abbreviations: t = top; b = bottom; c = centre; l = left; r = right)
© www.shutterstock.com:
1 c, 3 tl, 7 c, 9 t, 12 c, 15 cr, 16 b, 20 c, 21 tr, 21 bl, 25 tr, 26-27 c, 30 c, 31 r.
28 tr = © Solodov Aleksey / Shutterstock, Inc, 28 bl = © Holger Wulschlaeger / Shutterstock, Inc, 29 br © MISHELLA / Shutterstock, Inc

Wiki Commons:
13 bl Courtesy the NOAA Photo Library, 25 br © Hello32020

Nasa:
15 tr, 17 tl

2, cl = Deco / Alamy Stock Photo. 4, c = Ronnie Chua / Alamy Stock Photo. 6, c = Stocktrek Images, Inc. / Alamy Stock Photo. 8, c = REUTERS / Alamy Stock Photo. 10, c = Xinhua / Alamy Stock Photo. 11, cl = Xinhua / Alamy Stock Photo. 13, cl = Carol Lee / Alamy Stock Photo. 13, tr = epa european pressphoto agency b.v. / Alamy Stock Photo. 14/15, b = REUTERS / Alamy Stock Photo. 15, tr = NASA. 17, cl = NASA. 17, tr = RGB Ventures / SuperStock / Alamy Stock Photo. 18, c = Deco / Alamy Stock Photo. 22, c = ZUMA Press Inc / Alamy Stock Photo. 22, br = jay goebel / Alamy Stock Photo. 23, cl = epa european pressphoto agency b.v. / Alamy Stock Photo.23, br = Ryan McGinnis / Alamy Stock Photo. 24, c = Ryan McGinnis / Alamy Stock Photo. 25. Cl = Nature Photographers Ltd / Alamy Stock Photo. 27, br = William Brooks / Alamy Stock Photo. 29, tr = REUTERS / Alamy Stock Photo. 31, tr = Stocktrek Images, Inc. / Alamy Stock Photo

极端自然灾害

飓风与龙卷风

约翰·法恩登（英）著
陈泰之 译

外语教学与研究出版社
FOREIGN LANGUAGE TEACHING AND RESEARCH PRESS
北京 BEIJING

试想，面对9米高的水墙或者时速300千米的风会是怎样一种情景。当一个极端风暴来临时，危害和破坏必定也非常严重。这种突然释放的力量可以将房屋夷为平地，暴雨将淹没低洼地带，风暴将汽车像玩具一样抛向空中。

了解一下飓风和龙卷风最严重的后果，它们发生的原因和地点，还有用于追踪它们的无人机和卫星技术，以及勇敢地冲向危险的风暴追逐者……

HURRICANE
SEASON

目录

飓风

飓风是一个巨大的风暴，只有在空中才能看到它的全部。从高空俯瞰，飓风像一个正在旋转的巨型奶油蛋糕。白色部分是威力巨大的大片环状雷雨云，能够带来暴雨。风暴中心又称“风暴眼”，它周围有螺旋上升的强风，强风导致了风暴的旋转。

热带气旋

“飓风”是旋转上升的热带风暴。太平洋上类似的风暴也许也称为“台风”。科学家将飓风和台风统称为“热带气旋”。

龙卷风

龙卷风，也称陆龙卷、旋风，是一股急剧旋转、声大如雷的气流，气流直径只有几百米。龙卷风来自巨大的雷雨云，它横扫大地，常会摧毁建筑物、拔起大树、掀翻车辆。龙卷风外围的风速快得惊人，而它中心的气压却很低，这种气压差使得龙卷风中心具有强大的吸力，宛如一个巨大的真空吸尘器。

飓风是怎样形成的？

夏末，热带洋面受到强烈光照，形成大量水蒸气。水蒸气上升并逐步形成巨大的雷雨云，飓风随之产生。高空中的强风从东方吹来，掠过云层顶部，将云团聚合为一个大的螺旋式风暴。风暴向西旋转行进，将云团越聚越大，风暴的威力也随之增大。

你知道吗？

按照美国航空航天局的说法，一场飓风在其生命周期内释放的能量相当于10,000枚核弹爆炸产生的能量。

飓风的横截面

旋转的云团

在风暴眼的最中心，上升的气流再次下沉，带来了短暂的晴空和平静的天气

风暴眼

风从风暴眼里钻出

风在海平面上逆时针咆哮而过

剖析飓风

飓风发生时，雷雨的环状云中产生大量降雨，形成雨带。受海面强风的影响，风暴逆时针旋转。风暴正中间有一条穿透云层的通道，这条通道被称为“风暴眼”。风从风暴眼中盘旋上升，穿出云墙。

有力的旋转

一场大飓风一天时间内释放的能量相当于世界上所有发电站一年发出的能量。直径为数百千米的云环旋转运动，把行进中遇到的所有雷雨云聚集在一起，产生了这种能量。

雨带

来自非洲的风暴

大西洋上的飓风通常开始于非洲，近佛得角群岛。随后，它们壮大声势，在海面上以每小时24千米的速度向西移动。不到两周的时间，就能抵达加勒比海，然后向北行进。与此同时，飓风的能量也达到顶峰。

飓风的威力

飓风能带来超强大风。只有当风暴速度达到每小时118千米时，我们才能称其为“飓风”。在威力强大的飓风中，风力更加强劲。1969年发生的“卡米尔”飓风，风速达到了每小时305千米！

飓风的出生地

飓风通常在大西洋北部或太平洋东北部生成。飓风由此一路西进，在此过程中不断地积蓄能量，而后远离赤道，随着能量的耗尽逐渐消失。

登陆

飓风一旦登陆就会造成破坏。飓风登陆前，强风已经掀起巨浪，冲击海岸。岸上的观察者看到不祥的乌云正迎面袭来，感觉到风力逐渐抬升。他们即将大难临头了！

你知道吗？

2005年8月29日，飓风“卡特里娜”袭击了美国密西西比州的韦夫兰，风暴潮超过了7.9米。这不仅仅是一个海浪，它意味着整个海平面的高度超过了房子！

海地飓风

2007年，飓风“马修”袭击海地（见大图），强风吹折了当地所有的树，掀翻了几乎所有建筑物的屋顶，只有那些极为坚固的楼房得以幸免。它们也能轻易地吹走一辆大汽车。飓风带来的暴雨随即引发了洪水。

当心！

经过几个小时的蹂躏，风暴似乎消退了。风停雨住，或许还能见到太阳。但这只是风暴眼经过带来的短暂平息。用不了一个小时的时间，随着风暴眼继续前进，风雨会再次呼啸而来。

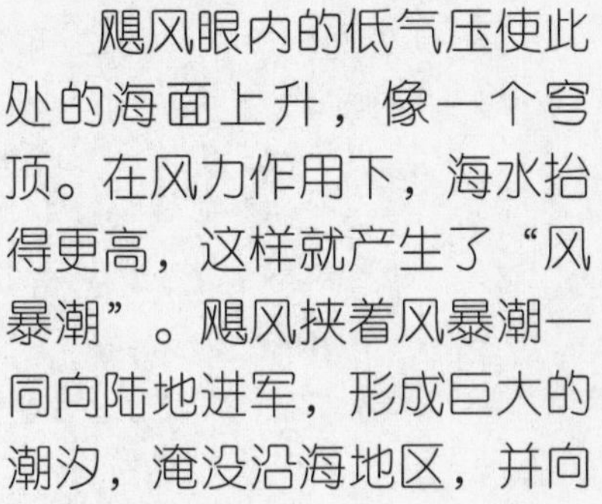

风暴潮

飓风眼内的低气压使此处的海面上升，像一个穹顶。在风力作用下，海水抬得更高，这样就产生了“风暴潮”。飓风挟着风暴潮一同向陆地进军，形成巨大的潮汐，淹没沿海地区，并向内陆漫延。

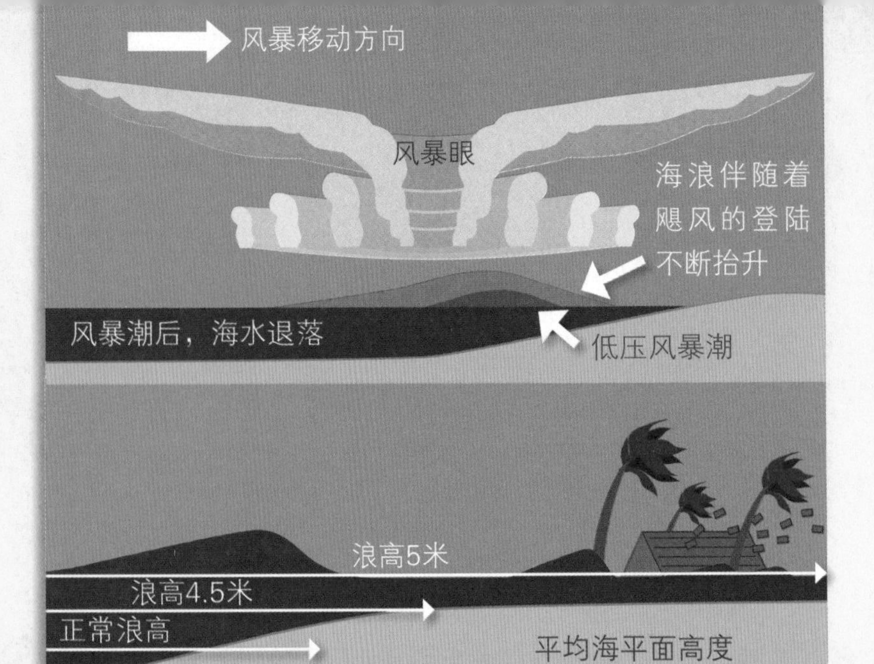

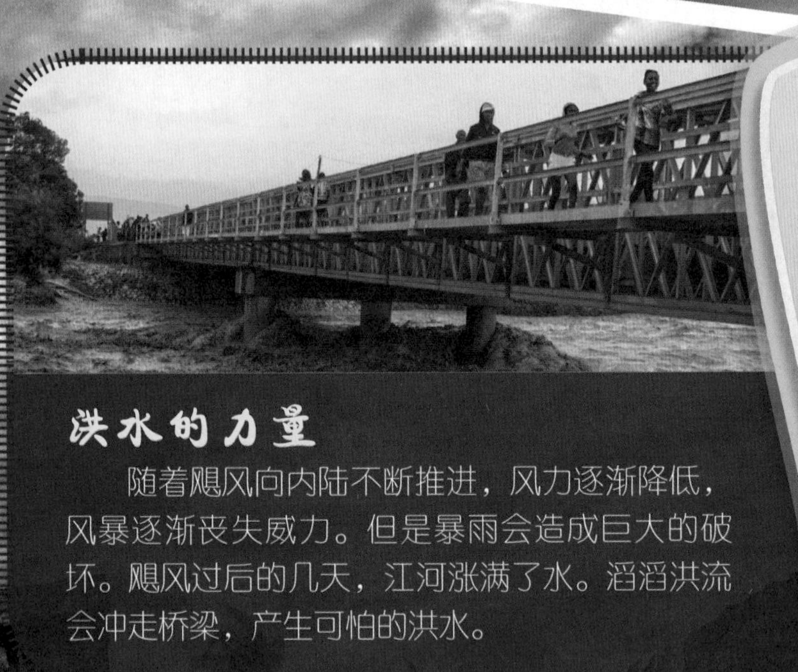

洪水的力量

随着飓风向内陆不断推进，风力逐渐降低，风暴逐渐丧失威力。但是暴雨会造成巨大的破坏。飓风过后的几天，江河涨满了水。滔滔洪流会冲走桥梁，产生可怕的洪水。

海浪的威力

遭受飓风侵袭的海岸会受到巨浪的重创。2004年9月15日，飓风“伊万”袭击美国的大西洋海岸时，掀起的海浪高达27米。想象一下10层楼高的水墙向你扑来的样子！

极端飓风的危害

有些飓风只是给人们的生活带来一些不便。而极端飓风却会造成巨大的破坏，特别是当它们袭击人口密集区时。狂风摧毁建筑物和输电线路。暴雨引发的洪水冲走桥梁，切断公路和铁路，使人们陷入困境。

你知道吗？

速度慢、强度小的飓风造成的破坏往往比速度快、强度大的飓风还要严重。因为这种飓风停留的时间长，造成暴雨和洪水的可能性更大。

魔鬼“桑迪”

2012年10月发生的飓风“桑迪”（见大图）是最大的飓风之一，直径宽达1,450千米。它的登陆给牙买加到美国的大范围地区造成了至少750亿美元的损失，这一数额仅次于飓风“卡特里娜”所造成的经济损失。

加尔维斯顿的灾难

1900年9月8日，有“南部纽约城”之称的美国得克萨斯州加尔维斯顿（见右图）被登陆的飓风完全摧毁。风速为每小时225千米的强风刮倒了很多建筑物，随即而来的4.5米高的风暴潮将废墟淹没。这次飓风造成6,000多人死亡，3,600座建筑物夷为废墟。

孟加拉国

油轮上岸

下图最能说明飓风“桑迪”的威力：一艘52米长的油轮冲上海岸，停在了纽约斯塔顿岛的前街上。这艘油船起初停泊在离海岸1.6千米以外的海湾上，但现在受风暴冲击上了岸。

恐怖的洪水

孟加拉国一半以上的区域海拔高度低于6米。更糟糕的是，孟加拉湾呈漏斗状，这意味着风暴潮会汹涌进入该国的低洼海岸线。因此，气旋给孟加拉国带来持久的洪水灾害，造成毁灭性的后果。

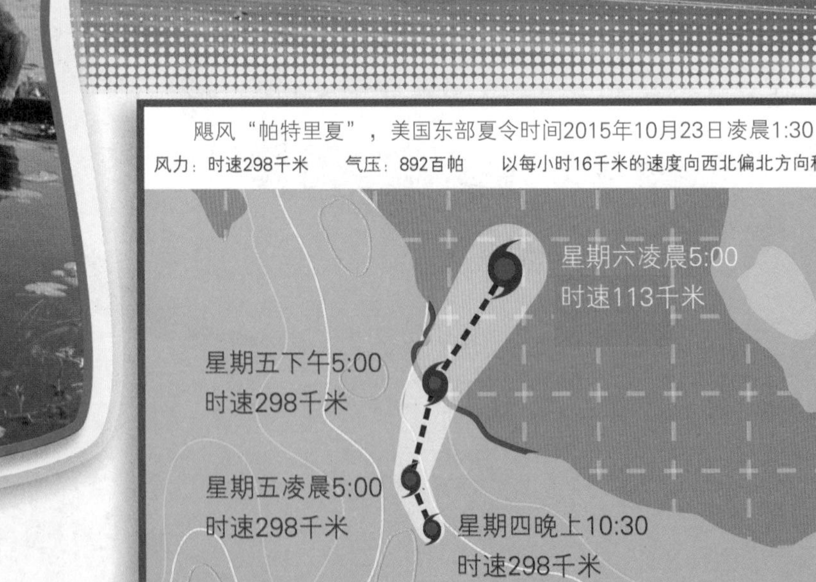

精力旺盛的“帕特里夏”

2015年10月发生的飓风“帕特里夏”是史上第二大飓风，其中心的低气压值在所有飓风中排名第二。它在墨西哥登陆，风力强劲。幸运的是，“帕特里夏”径直穿越山区和乡间，避开了城市，所以没有造成太大的损失。

极端飓风的故事

生活在美国路易斯安那州新奥尔良市的人永远不会忘记2005年8月发生的飓风“卡特里娜”。这是美国历史上最致命的飓风之一，当然也是造成损失最大的自然灾害。灾害造成超过1,800人死亡，估计造成的经济损失超过1600亿美元。

你知道吗？

飓风“卡特里娜”引发的风暴潮高达5到9米，这是美国历史上记录到的最大高度。

决堤

新奥尔良市地势低洼，需要建立“防洪堤”来抵御洪水。但在“卡特里娜”来临期间，受风暴潮和暴雨影响，河水暴涨。很快水就溢出岸堤，防洪堤出现了50处缺口，洪水涌进城市。

破坏

右边的两幅图片是新奥尔良沿海地区的一处房屋遭遇飓风“卡特里娜”前后的情形。该飓风总共摧毁100多万座房屋，其中仅新奥尔良就有134,000座，几乎占据了该市全部房屋的四分之三。房屋受损的主要原因是洪水。

热量的威力

“卡特里娜”抵达墨西哥湾时还是一场温和的飓风，但是那里异常温暖的海水急剧增加了飓风的威力。短短9个小时内，风暴的规模翻了一番，飓风强度从3级升至5级（等级描述详见第17页）。“卡特里娜”是迄今为止袭击过美国的四个5级飓风之一。

“大快活”

当飓风“卡特里娜”袭来时，新奥尔良（别名是“大快活”，译者注）是一个繁华的城市，有近500,000居民。但是，随着飓风逼近，许多人撤离了，全市80%的地区被洪水淹没。一年之后只有200,000人返回，让这座城市失去了往日的喧嚣。

飓风发生前的预警

每年的6—11月，大西洋上会发生约10场飓风，其中一场或两场飓风将袭击美国海岸。这是飓风多发季节，位于佛罗里达州的美国国家飓风中心高度戒备，密切关注飓风的发展动向，尽可能多地给人们预警。

你知道吗？

飓风猎人飞机（见右图）有时会穿过飓风的眼墙——这是风暴威力最大的位置！飞机遭受暴雨和冰雹击打，被时速240千米或者更猛烈的风吹打，在强烈的升降气流中颠簸。

空中鸟瞰

科学家借助高空的卫星可以轻松地跟踪飓风的移动路径。通过把连续几个小时的卫星照片放在一起，就能看出飓风的发展情况。但是，飓风也会迅速改变路径。上图为2016年10月飓风“马修”逼近佛罗里达州的情形。

飓风猎人

无人机或机器人飞机可以通过遥控进入飓风内部。无人机可以用来监测飓风最危险区域的情况，同时又不会危及人员的生命。它们能在两倍于客机的高度上飞行30小时。

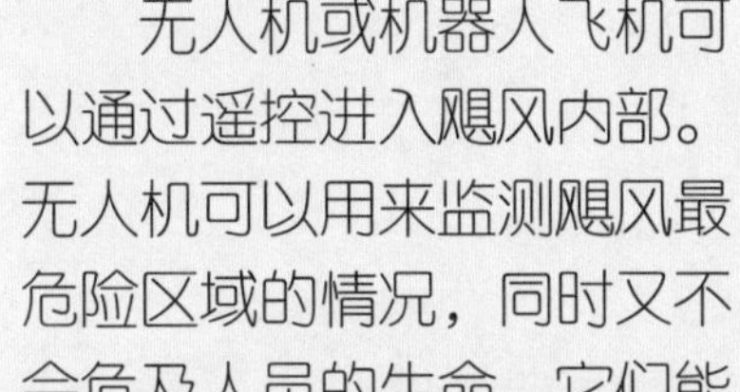

核实降雨情况

多普勒雷达借助微波的反射来探测物体。利用特制的多普勒雷达，飓风观察员能够制作出详尽的地图，标明降雨的区域和雨量，甚至可以探测风速。该图中的雷达站坐落于美国堪萨斯州的道奇城。

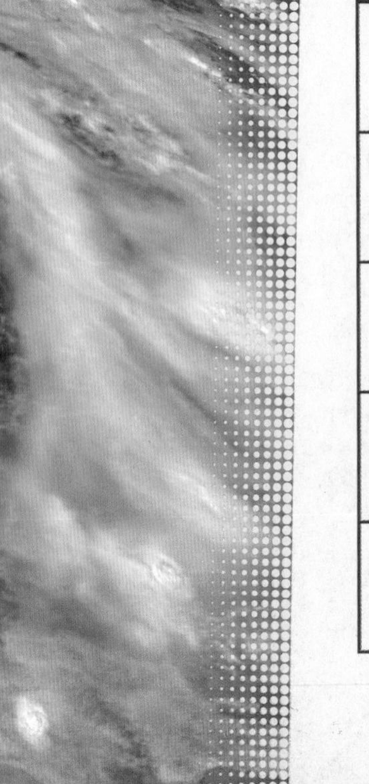

飓风等级

飓风的力量按照赛福尔-辛普森飓风等级分为1—5级。

等级	风速	破坏程度
1	119—153千米/小时	轻度破坏：活动房屋轻微挪动；广告牌吹倒；树枝折断
2	154—177千米/小时	中度破坏：活动房屋被掀翻；屋顶被掀开
3	178—208千米/小时	大规模破坏：小型建筑物倒塌；树木连根拔起
4	209—251千米/小时	极度破坏：大多数树木被刮倒；建筑物受到普遍的结构性破坏
5	252千米/小时以上	灾难性破坏：大多数建筑物被毁，森林、道路和管道被毁

龙卷风是怎样形成的？

横跨北美大平原的“龙卷通道”是世界上最危险的龙卷风多发地带。来自墨西哥湾的暖空气与落基山脉吹来的冷风在此相遇，形成号称“超级单体”的巨大雷暴，这些雷暴造就了威力强大的龙卷风。

你知道吗？

通常，龙卷风的旋转风速约为每小时160千米，但有时能够达到每小时480千米，风力能够将一根圆木头卷起并用其砸透砖墙！

雷鸣与闪电

龙卷风来临时十分可怕。咆哮的旋风犹如一列呼啸而来的货运列车，雷声隆隆，电光闪闪，从云端霹雳直下。幸好大概10分钟的时间它们就逐渐消失了。

龙卷通道

“龙卷通道”是美国中部这段龙卷风多发区域的俗称。来自北方的冷空气和来自南方的暖空气在这里交汇，导致该地区成为龙卷风多发区域。得克萨斯州每年发生的龙卷风次数最多（这也跟它的州面积太大有关），堪萨斯州、俄克拉何马州和佛罗里达州也经常遭受龙卷风袭击。

龙卷风“制造者”

当“超级单体”内强烈的上升气流遭遇云团上方的冷风时，麻烦就产生了。两种气流的冲撞使得上升气流发生旋转，制造出转动的气柱，也叫中气旋。伴随着风暴的加剧和降雨，气柱呈现剧烈旋转的漏斗形，从云团底部下落。龙卷风就形成了。

极端龙卷风的危害

通常，除了晃动树木或吹落松动的门，旋转的龙卷风造成的破坏不大。但有时龙卷风会带来灾难性的危害，将建筑物夷为平地，把汽车抛到空中，造成人畜死伤。根据改良藤田级数，龙卷风的强度等级可划分为从0到5的六个等级。

你知道吗？

美国历史上最致命的龙卷风发生在1925年3月18日。这场龙卷风夺去了695人的生命，穿越了密苏里州、伊利诺伊州和印第安纳州，行程480多千米。根据藤田级数，此次龙卷风被定为5级龙卷风。

房屋倒塌

密苏里州乔普林市的一处房屋在2011年5月的一场5级龙卷风中被摧毁（见右图），这足以表明龙卷风的破坏力。奇怪的是，书籍还整齐地摆放在书架上，这说明龙卷风的威力具有局部性的特点。

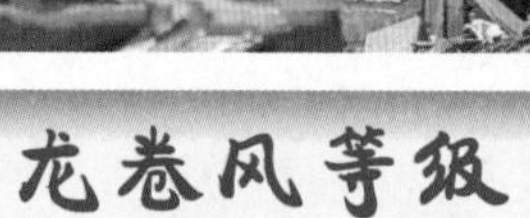

骇人听闻：青蛙雨

令人惊奇的是，暴风雨时天空不仅下雨，还经常下青蛙！理论上来说，青蛙是在龙卷风掠过池塘时被卷走的。水面上形成的龙卷风，也称为“水龙卷”，甚至能把鱼卷走！

龙卷风等级

根据改良藤田级数，龙卷风造成的破坏可以分为0—5级。

等级	风速	破坏程度
EF0	105—137千米/小时	轻度破坏：屋顶瓦片吹动；树枝折断
EF1	138—177千米/小时	中度破坏：屋顶瓦片吹落；活动房屋倾覆；汽车吹离道路
EF2	178—217千米/小时	大范围破坏：屋顶吹落；活动房屋被毁；铁路上的运货车皮翻倒；大树连根拔起
EF3	218—266千米/小时	严重破坏：房屋吹倒；汽车抛向空中；重物像导弹发射一样投射出去
EF4	267—322千米/小时	极端破坏：即使是地基坚固的房屋也被毁；汽车变成飞行的导弹；树皮剥离树干
EF5	322千米/小时以上	毁灭破坏：灾难性的、大范围的破坏

极端龙卷风的故事

2011年5月22日下午，美国历史上最凶猛的龙卷风之一袭击了密苏里州乔普林市。龙卷风袭来的24分钟前警报声响起，但许多人未予理会，毕竟之前有过太多的假警报。可是，这一次却是真的。

你知道吗？

龙卷风对乔普林市的突然袭击造成了毁灭性破坏，人们震惊不已。一个满脸茫然的受灾市民走向一名救援人员，问："他们什么时候给我的房子通电？"而他的房子已经被完全摧毁了。

夷为平地

时速400千米的强风掠过城镇，给城镇造成了严重损失，这意味着这场龙卷风有可能达到了EF5级。在被夷为平地的医院，人们发现了一辆大卡车的残骸。这辆卡车被风移动了115米的距离。卡车残骸挂在一棵被剥去树皮的树上。

灾难前后

龙卷风袭击之前，乔普林市是一座有50,000人口的宁静小镇，这里有许多年逾百岁的坚固砖房。但是在龙卷风经过的几分钟后，多达7,000幢房屋成为瓦砾，就像经历了原子弹爆炸一样。

风暴的踪迹

途径乔普林市的龙卷风划出了一条宽度约1600米的破坏路径。参与重建工作的扎克·罗森堡说："龙卷风的轨迹清晰明了。有的街道一侧被彻底摧毁，而它另一侧则完好无损。"

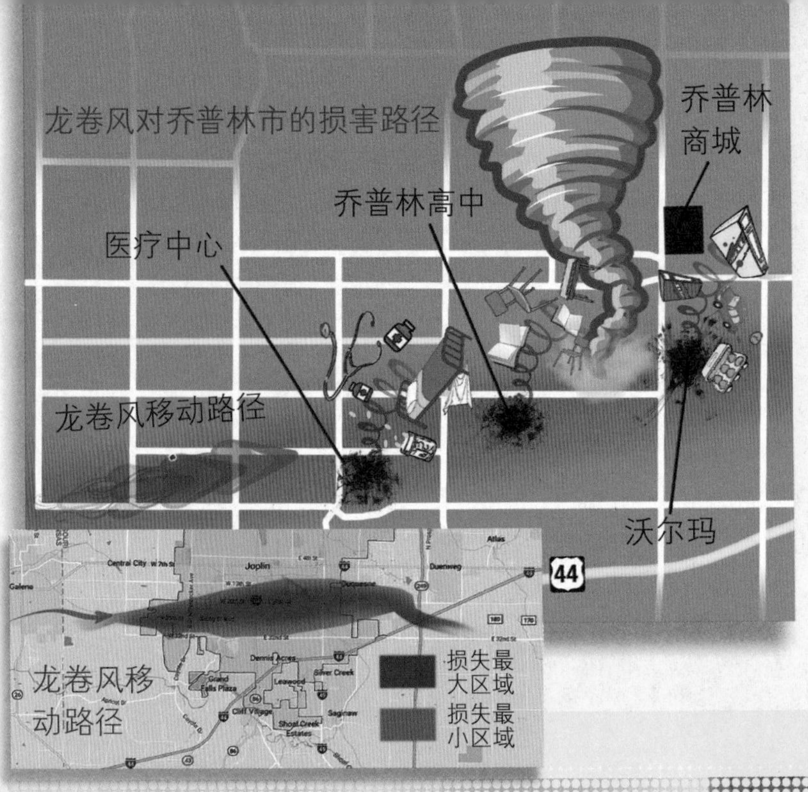

灾后重建

89岁的唐·阿特伯里是乔普林市房屋被毁的居民之一（见上图）。后来，城市开展了成效显著的重建工程，到2016年这里很难看出龙卷风遗留的痕迹了。唐·阿特伯里的房子也焕然一新。

极度破坏

乔普林市的龙卷风是美国历史上第七大致命的龙卷风，造成161人死亡。这场仅仅持续数分钟的灾难破坏力巨大，大多数专家将其定级为EF5级龙卷风。卡车和水泥砖块被抛到数百米以外，这表明风速超过了每小时320千米。

龙卷风发生前的预警

在过去的半个世纪中，我们对龙卷风的认识有了长足的进展，这要部分得益于勇敢的风暴追逐者。一些风暴追逐者是摄影师，他们希望捕捉到令人惊叹的镜头。其他人，如“涡旋2”和“旋转2012”项目的科学家，则在探寻诱发龙卷风的相关条件数据。

目击者

“我一闭上眼睛还能听见那个声响。它让我永远忘不了——像同时出现了呼啸般的尖锐噪声和低沉的隆隆声。”风暴追逐者史蒂夫·约翰逊这样描述龙卷风逼近的情景。

鸟能觉察到龙卷风的来临吗？

2014年科学家发现，他们正在追踪的五只小金翅虫森莺突然逃离了巢穴。隔天，龙卷风来袭。几天之后，龙卷风过去了，鸟儿安全返回巢穴。科学家相信，借助超灵敏的听觉，它们能听到远处龙卷风的声音。

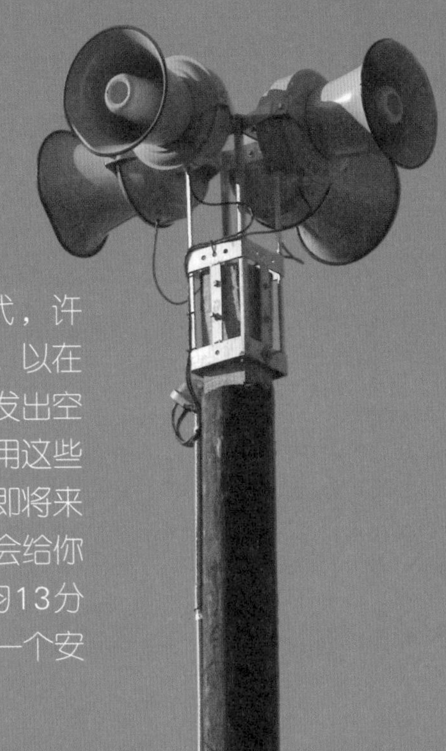

龙卷风警报

在20世纪50年代，许多地方设置了警报器，以在危险来临之时向人们发出空袭警报。现在，人们用这些警报器来警告龙卷风即将来临。请注意，它们不会给你留下太多时间：平均13分钟。但这足够你去到一个安全的地方了！

龙卷风之声

多普勒雷达在追踪龙卷风的问题上至关重要。它们同常规雷达一样，通过发送无线电波和观察电波反弹来探测事物。但是多普勒雷达还可以根据风暴云中水滴反弹的无线电波频率的变化，探知风暴移动的方式，以及它的位置。

龙卷风追逐者

美国电影制作人肖恩·凯西因其在拍摄电视剧《风暴追逐者》中的英勇行为而闻名。为了足够接近龙卷风，他建造了有两块装甲板的“龙卷风截听车”（简称TIVs），上面配备了摄像机转塔、天气传感器、防弹窗，还安装了下拉金属挡板来保护车辆的底盘，防止龙卷风夹带的投射物的冲撞。

严峻的未来

本世纪已经发生了多次极端风暴，从飓风“卡特里娜”到热带风暴“纳尔吉斯”，后者于2008年在缅甸造成138，000人死亡。然而，许多科学家认为，我们将面临更加严重的风暴。由污染气体（如二氧化碳）导致的世界性气候变化可能给大气层增加更多的能量，从而破坏天气模式。

不断增长的代价

美国有记录的10个代价最为惨重的飓风中，有7个发生在2004年之后。飓风“卡特里娜”（2005年）造成损失1,600亿美元，飓风“桑迪”（2012年）造成损失750亿美元。飓风造成的损失越来越严重，一方面是由于沿海地区的居住人口越来越多，另一方面是因为风暴自身似乎也变得更加凶猛。

龙卷风群

过去的10年间，科学家们注意到龙卷风的模式发生了变化。虽然龙卷风发生的次数没有很大变化，但龙卷风现在似乎是在毁灭性地成群出现，相互之间保持较大间隔。2011年发生在乔普林市的龙卷风正是这样一个风群带来的。一些科学家认为龙卷风群的出现应归咎于气候变化。

风暴天气

飓风发生得越来越频繁。1966年至2009年间，平均每年发生6次充分发展的飓风。自2009年以来，每年发生8次飓风。这是因为海洋变得更温暖了。科学家们尚未发现海洋变暖的确切原因，但许多科学家相信这是气候变化导致的。

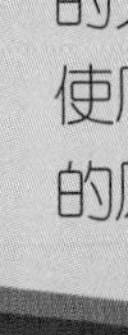

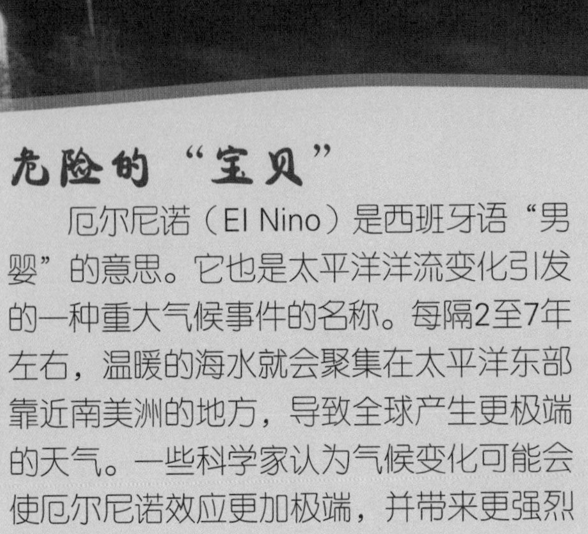

危险的“宝贝”

厄尔尼诺（El Nino）是西班牙语“男婴”的意思。它也是太平洋洋流变化引发的一种重大气候事件的名称。每隔2至7年左右，温暖的海水就会聚集在太平洋东部靠近南美洲的地方，导致全球产生更极端的天气。一些科学家认为气候变化可能会使厄尔尼诺效应更加极端，并带来更强烈的风暴。

大事件时间轴

1054年

欧洲最早记录的龙卷风袭击了爱尔兰吉尔贝根附近的罗斯达拉

1281年

日本博多湾的台风摧毁了整个元军舰队，被日军称为“神风”

1609年

在百慕大的一场飓风中，殖民者的“海洋探险”号（见左图）船只失事。以此为灵感，莎士比亚创作出名剧《暴风雨》

1635年

“殖民地”大飓风是已知的第一场发生在美国新英格兰地区的飓风

1780年

大西洋上最致命的飓风之一“大飓风”夺去了加勒比地区20,000多人的生命

1815年

“九月”大风袭击新英格兰地区

1054

1494年

欧洲的首次飓风记录由哥伦布完成。他率先乘坐“圣玛丽亚”号（图为“圣玛利亚”号复制品）驶往了美洲

1559年

飓风摧毁了佛罗里达的第一个殖民地

1667年

美国弗吉尼亚遭遇了一场当地最严重的飓风，飓风摧毁了10,000座房屋

1737年

袭击印度和孟加拉的“胡格利河”气旋是有史以来最致命的风暴之一，造成350,000多人死亡，损毁船只20,000条

1839年

“科林加”气旋袭击了印度东南部，造成毁灭性灾难，近300,000人在灾害中遇难

1881年

“海防”台风途径中国南海北部湾，产生巨大的风暴潮，造成300,000人遇难

1974年

24小时内，美国出现了148个龙卷风，形成了一次龙卷风的超级大爆发

1925年

袭击了密苏里州、伊利诺伊州和印第安纳州的“三州”龙卷风是已知最大、威力最强的龙卷风之一

1923年

一场大地震引发了台风及火灾，台风助长了火势，摧毁了东京

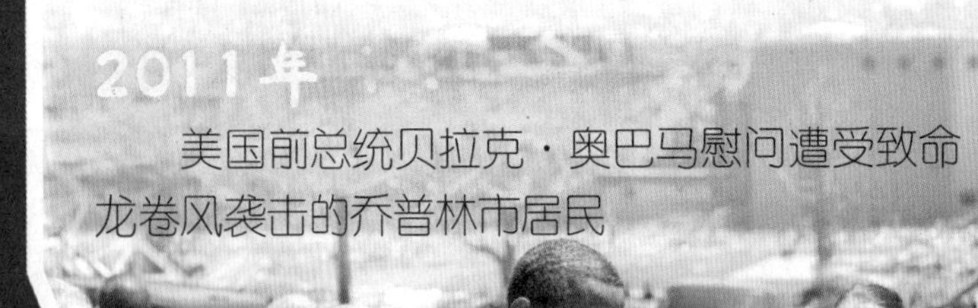

2011年

美国前总统贝拉克·奥巴马慰问遭受致命龙卷风袭击的乔普林市居民

1992年

飓风“安德鲁”是美国历史上造成经济损失排名第三的飓风，带来265亿美元的损失

1988年

飓风“吉尔伯特”是有史以来最剧烈的风暴之一，给中美洲造成严重损失

2005年

飓风“卡特里娜”是美国造成经济损失最大的飓风，摧毁了新奥尔良市

1970年

“波拉”大气旋袭击孟加拉，造成500,000人死亡，成为危害最大的自然灾害之一

1989年

史上最致命的龙卷风袭击了孟加拉道拉特布尔-萨图里亚地区，夺去1,300人的生命

1975年

超级台风“尼娜”摧毁了中国的数个大坝，洪水迫使11,000,000人离开家园

1900年

发生在得克萨斯州的“加尔维斯顿”飓风是袭击美国的最强烈飓风之一，飓风带来的风暴潮淹没了整个加尔维斯顿城

2010年

飓风“桑迪”袭击美国东南海岸，一直席卷到纽约市（上图为“桑迪”过后的情形）

不可思议！

关于飓风和龙卷风的惊人事实

搭便车

有传闻说，美国俄克拉何马州的一小群牛被龙卷风吸起来，然后在另外一个地方安然无恙地落地。这可能不是真的。但是在2011年4月，据说一个男孩从床上被风卷走，扔到了离家几百米的地方。这种幸免于难的经历非常罕见。

最大的风暴

地球上最大和最强的飓风直径约达1,600千米，风的时速达320千米，非常可怕。但在木星上，有一场风暴的直径是整个地球的两倍，风速超过每小时640千米。而且，它至今至少持续了150年，人们称它为“大红斑”。

为风暴命名！

每一个飓风、气旋和台风都有自己的名字，以便于预报员对其进行追踪。20世纪初，一位名叫克莱门特·雷格的澳大利亚气象预报员最先为飓风命名。每年第一个飓风的名字首字母为A。如果一个飓风破坏力极大，它的名字会永久退隐，另一个名字会取而代之。

错误做法！

有些人会告诉你，应该打开窗户让龙卷风安全通过。这是错误的做法。开错窗户可能使气流进入室内，从室内把房屋摧毁。

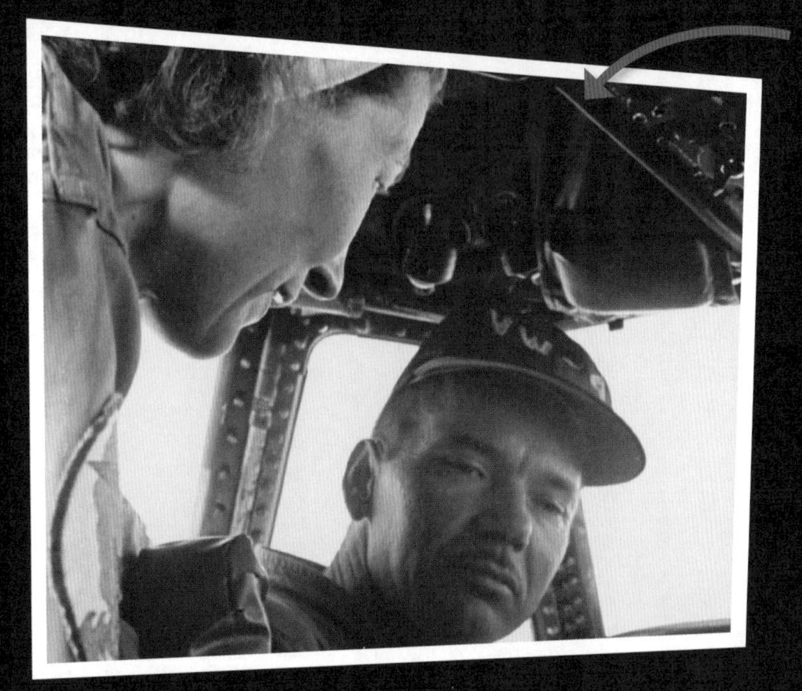

人工影响飓风计划

从1962年到1983年，美国政府启动人工影响飓风计划，利用飞越飓风上空的飞机把碘化银粉末投放入云团中，力图催生雨滴以冷却飓风。但这个计划没有奏效。

恶魔

龙卷风一词出自西班牙语tornada，意思是“大雷雨”。飓风一词出自北美土著人泰诺人的词汇hurucane，意思是“风的恶魔”。

气旋之最

全球死亡人数最多
“波拉”大气旋（孟加拉）
时间：1970年11月12日
死亡人数：多达500,000人

损失最大
飓风“卡特里娜”（美国）
时间：2005年8月29至31日
死亡人数：1,836人
经济损失：1,600亿美元

风力最强
“奥利维娅”气旋（澳大利亚）
时间：1996年4月10日
风速：每小时408千米

降雨量最大
“海厄森斯”气旋（印度洋）
时间：1980年1月14至28日
降雨量：6,083毫米

龙卷风之最

全球死亡人数最多
道拉特布尔-萨图里亚（孟加拉）
时间：1989年4月26日
死亡人数：1,300人

美国死亡人数最多
“三州”龙卷风（美国）
时间：1925年3月18日
死亡人数：695人

损失最大
2011年“超级大爆发”（美国）
时间：2011年4月26至28日
死亡人数：348人
经济损失：110亿美元

有记录以来最大
埃尔雷诺（美国）
时间：2013年5月31日
直径：4.2千米

索引

作者简介

约翰·法恩登，英国伦敦城市行业协会的皇家文学基金会研究员。他为成人与儿童写作了大量图书，涉及科学、技术、自然等领域，包括畅销世界的《你觉得自己聪明吗？》和《放我出去》。他六次入围英国皇家学会青少年科学图书奖，入围书目有《地球原理》《何因何果》和《人体项目》（2016）等。

Picture Credits (abbreviations: t = top; b = bottom; c = centre; l = left; r = right)
© www.shutterstock.com:
1 c, 3 tl, 7 c, 9 t, 12 c, 15 cr, 16 b, 20 c, 21 tr, 21 bl, 25 tr, 26-27 c, 30 c, 31 r.
28 tr = © Solodov Aleksey / Shutterstock, Inc, 28 bl = © Holger Wulschlaeger / Shutterstock, Inc, 29 br © MISHELLA / Shutterstock, Inc

Wiki Commons:
13 bl Courtesy the NOAA Photo Library, 25 br © Hello32020

Nasa:
15 tr, 17 tl

2, cl = Deco / Alamy Stock Photo. 4, c = Ronnie Chua / Alamy Stock Photo. 6, c = Stocktrek Images, Inc. / Alamy Stock Photo. 8, c = REUTERS / Alamy Stock Photo. 10, c = Xinhua / Alamy Stock Photo. 11, cl = Xinhua / Alamy Stock Photo. 13, cl = Carol Lee / Alamy Stock Photo. 13, tr = epa european pressphoto agency b.v. / Alamy Stock Photo. 14/15, b = REUTERS / Alamy Stock Photo. 15, tr = NASA. 17, cl = NASA. 17, tr = RGB Ventures / SuperStock / Alamy Stock Photo. 18, c = Deco / Alamy Stock Photo. 22, c = ZUMA Press Inc / Alamy Stock Photo. 22, br = jay goebel / Alamy Stock Photo. 23, cl = epa european pressphoto agency b.v. / Alamy Stock Photo.23, br = Ryan McGinnis / Alamy Stock Photo. 24, c = Ryan McGinnis / Alamy Stock Photo. 25. Cl = Nature Photographers Ltd / Alamy Stock Photo. 27, br = William Brooks / Alamy Stock Photo. 29, tr = REUTERS / Alamy Stock Photo. 31, tr = Stocktrek Images, Inc. / Alamy Stock Photo

WHEN DISASTER STRIKES
EXTREME EARTHQUAKES AND TSUNAMIS

Written by John Farndon

外语教学与研究出版社
FOREIGN LANGUAGE TEACHING AND RESEARCH PRESS
北京　BEIJING

An earthquake can shift an entire city three metres, make the Earth spin faster, and send shudders right across the world. It is raw power at work – and then monster waves follow... These may travel thousands of kilometres across the ocean, dump ships on dry land, and ruin a million buildings in one fatal whoosh.

Find out about the most fearsome earthquakes and tsunamis, their causes and devastating effects, and the latest technology used to detect these hidden terrors. And virtually nowhere on the planet is entirely safe from them...

CONTENTS

EARTHQUAKES

Earthquakes are a violent shaking of the ground. They are set off by a sudden snapping or shifting of the giant slabs of rock that make up the Earth's surface. Most are so faint that they are only detectable on the most sensitive equipment. But a few are so powerful they can cause devastation, destroying cities and killing many people.

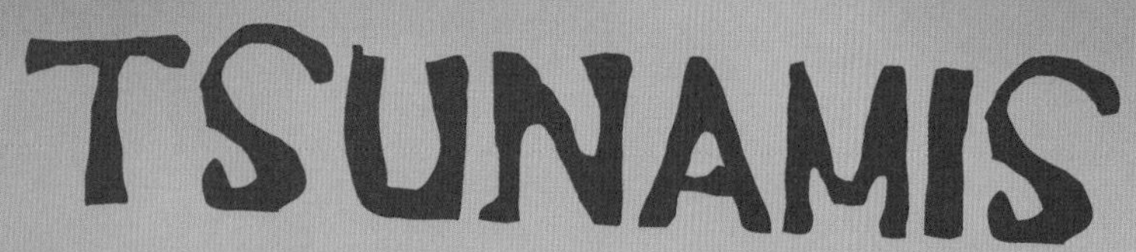

TSUNAMIS

A tsunami is a wave, or series of waves, set off when an earthquake or volcanic eruption under the sea moves a lot of water. The movement spreads out in all directions. While under the sea, it cannot be seen, but once it reaches the coast it sends up huge walls of water that can cause devastation, washing away everything in its path.

HOW DOES AN EARTHQUAKE HAPPEN?

The Earth's solid surface, or crust, is cracked into giant slabs of rock called tectonic plates. These slabs are always grinding past each other. Most earthquakes happen when two slabs suddenly crack or slip. As they slip, they send out shock waves or 'seismic waves' through the ground.

UNBELIEVABLE!

There is no chance of running away from an earthquake. The fastest earthquake waves roar through the ground at 8 km/sec – that's over 40 times as fast as a jet airliner!

CALIFORNIA ALERT

The San Andreas Fault zone in California in the USA is a transform (right) over 1,300 km long. Here the tectonic plate under the Pacific Ocean grinds past the North American plate. Movement along this fault frequently causes earthquakes, such as the quake that devastated San Francisco in 1906. Experts believe it will cause another massive one in the future.

SHAKING PLATES

Some tectonic plates pull apart, some push together, and some slide sideways past each other. It is these sliding junctions, known as 'transforms', that set off most earthquakes. The jagged edges of the plates snag together, allowing pressure to build up, until they suddenly slip, setting off an earthquake.

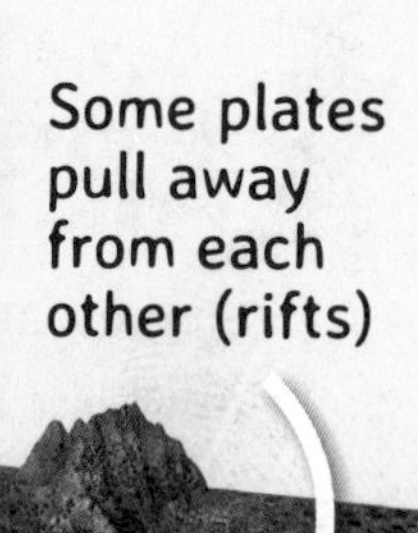

Some plates slide sideways past each other (transforms)

Some plates push on top of another (subduction)

UNDERWATER VOLCANO

Most earthquakes are set off by tectonic plates juddering past each other. But a few may be set off by volcanoes. As the hot magma pushes into cracks in the rock under a volcano, it can make the rock snap and send out an earthquake.

QUAKE ZONES

Nowhere in the world is completely immune from earthquakes. Yet most big earthquakes occur in belts or 'earthquake zones' along the boundaries between tectonic plates. Around 80 per cent of all big quakes strike around the edge of the Pacific Ocean, while many of the rest rumble through Southern Europe, the Near East and Southern Asia.

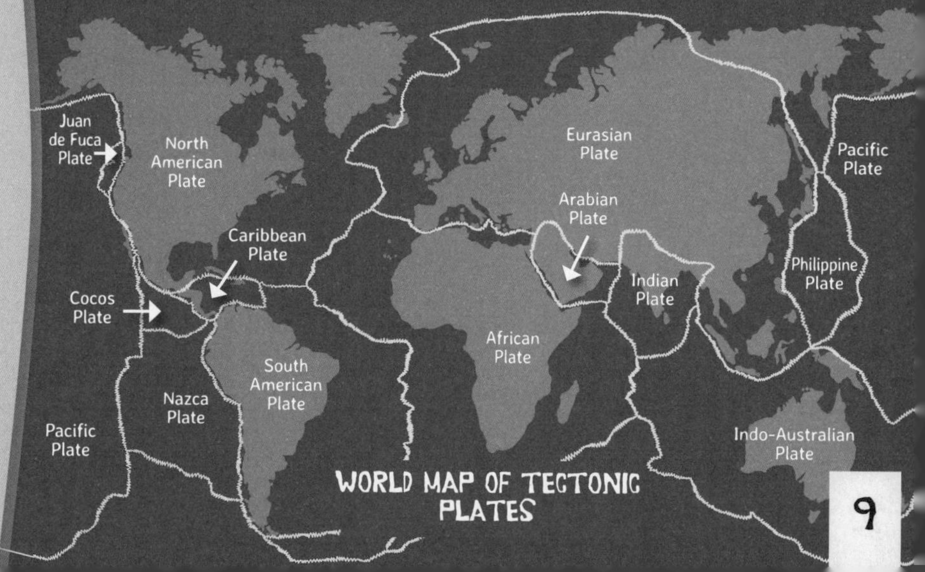

WORLD MAP OF TECTONIC PLATES

EARTHQUAKE DAMAGE

When an earthquake hits, it shakes the ground violently to and fro. At the epicentre of a powerful shallow quake, the shaking is so severe that even the strongest buildings may not survive. It can also open up giant cracks, and throw the ground sideways or up or down, or start avalanches. It can also turn the ground almost to liquid.

UNBELIEVABLE!

In 2010, an earthquake in Chile moved the city of Concepción 3 m to the west. The quake also shortened Earth's day fractionally!

HAITI TRAGEDY

The quake that hit the island republic of Haiti and its capital Port-au-Prince (main picture) on 12 January 2010 had a terrible effect. The poorly built houses collapsed, over 316,000 people died, and 3,000,000 had their homes ruined. Disruption of water supplies led to a horrible outbreak of the disease cholera.

FIRE DAMAGE

Some of the worst earthquake damage is caused by fire, often set off by damage to gas pipes and electrical cables. When a huge earthquake hit San Francisco on 18 April 1906, it was one of the worst natural disasters in US history. But most of the damage was done by the fires that raged for three days after the quake.

DEADLY SURPRISE

One of the worst aspects of earthquakes is that they can strike out of the blue. Even in an earthquake zone, centuries can go by with no disturbance. That's what happened in the historic town of Amatrice in Italy. Then with no warning, in August 2016, the town was utterly destroyed by a quake in just a few minutes.

BROKEN BRIDGE

Bridges are very vulnerable to quakes. Just a tiny movement can snap a span, with devastating consequences. So engineers in quake zones build bridges such as the San Francisco Bay Bridge, to be as earthquake-resistant as possible.

AFTER AN EXTREME EARTHQUAKE

Few earthquakes last for more than a minute. But in that brief time, they can do terrible damage. The sooner the emergency services can move in to help, the better. But with roads broken, water and power supplies cut off, and buildings in danger of further collapse, they have a very challenging task.

UNBELIEVABLE!

A massive $3.5 billion was raised by ordinary people worldwide to help victims of the 2010 Haiti earthquake. Some movie stars donated $1 million each, while Brazilian model Gisele Bundchen gave $1.5 million and Tiger Woods $3 million.

REBUILDING

Rebuilding after an earthquake is a long business, especially in poor countries where there are few resources. Even two years after the 2010 Haiti earthquake, half a million people were still homeless, and cholera was making many ill.

DETECTION EQUIPMENT

Modern technology can be a great help in finding survivors. Video cameras can be squeezed through holes on narrow poles. Thermal imagers can pick up body heat in dark places. And specialist sound equipment can pick up the sounds of breathing — but rescuers have to be very quiet themselves.

RESCUE DOGS

One of the first tasks is to rescue survivors still trapped under collapsed buildings. To help find them, rescuers may use sniffer dogs whose keen sense of smell enables them to pick up on signs of life that human rescuers cannot. The dogs can also work over a large area quickly.

OUTSIDE HELP

Victims of earthquakes need help not only with rescue efforts but rebuilding afterwards. Disaster relief workers, rescue teams, medical staff, technicians and security personnel may be sent to help at once. And ordinary people around the world may raise money to buy everything from fresh water to new furniture.

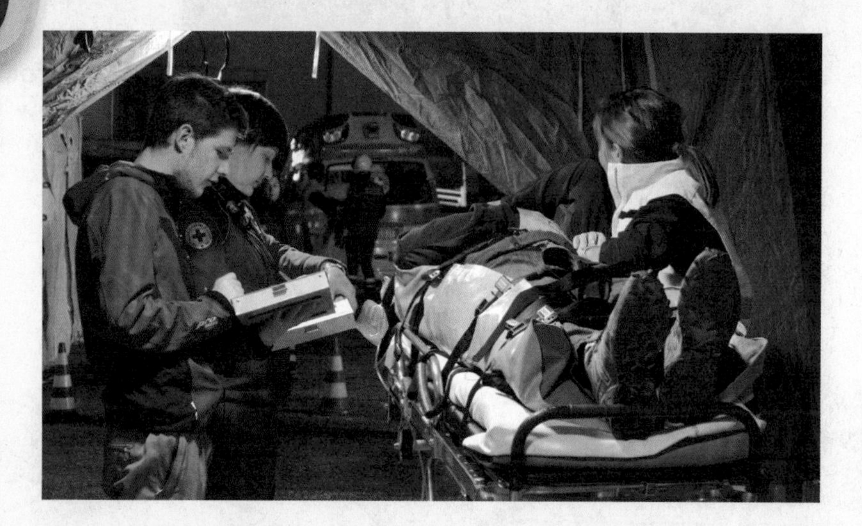

MEDICAL AID

Doctors and nurses are vital in the aftermath of a quake, not only for treating terrible injuries. Illnesses may set in as food and water supplies are disrupted. Water often gets polluted by dirt and sewage, causing cholera, dysentery and other water-borne diseases.

EXTREME EARTHQUAKE STORY

The earthquake that hit the town of Valdivia in Chile on 22 May 1960 was the most powerful ever recorded. It began 160 km offshore at 3:11 pm and shook Valdivia and towns along the coast almost instantly. But it was the giant tsunami that hit the shore 15 minutes later that did most of the damage.

UNBELIEVABLE!

A quarter of all the earthquake energy of the 20th century, and of the 2004 Indian Ocean quake, was concentrated in the 1960 Chile quake.

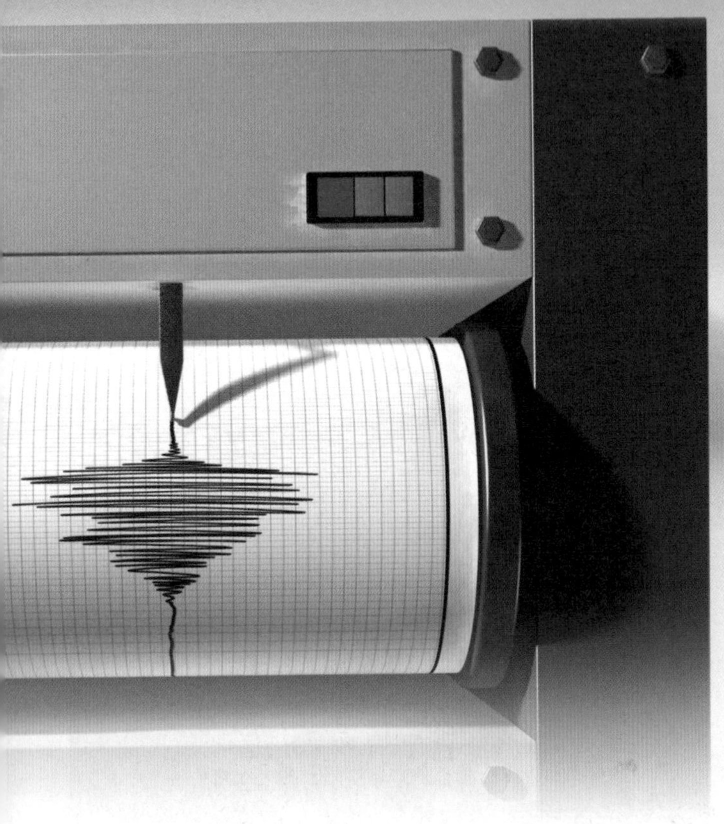

HOW BIG?

Scientists use devices called seismometers to register an earthquake's vibrations or waves. In the past, they rated their size on a scale, called the Richter scale, from 0 to over 9, the strongest. But for the biggest earthquakes, they now use the Moment Magnitude scale. This combines Richter readings with observations of rock movements to show the true power of an earthquake.

Plate movement

Seismic waves

Focus

Plate movement

POWER QUAKES

The Valdivia quake was 9.4–9.6 on the Moment Magnitude scale. Only six other quakes have ever been above 9.

22 May 1960	Valdivia, Chile	9.4–9.6
27 March 1964	Prince William Sound, Alaska,USA	9.2
26 December 2004	Sumatra, Indonesia	9.1–9.3
11 March 2011	Tōhoku region, Japan	9.1[4]
4 November 1952	Kamchatka, USSR	9.0
13 August 1868	Arica, Chile (then Peru)	9.0
26 January 1700	Cascadia, Pacific Ocean	8.7–9.2

1960 Valdivia Earthquake

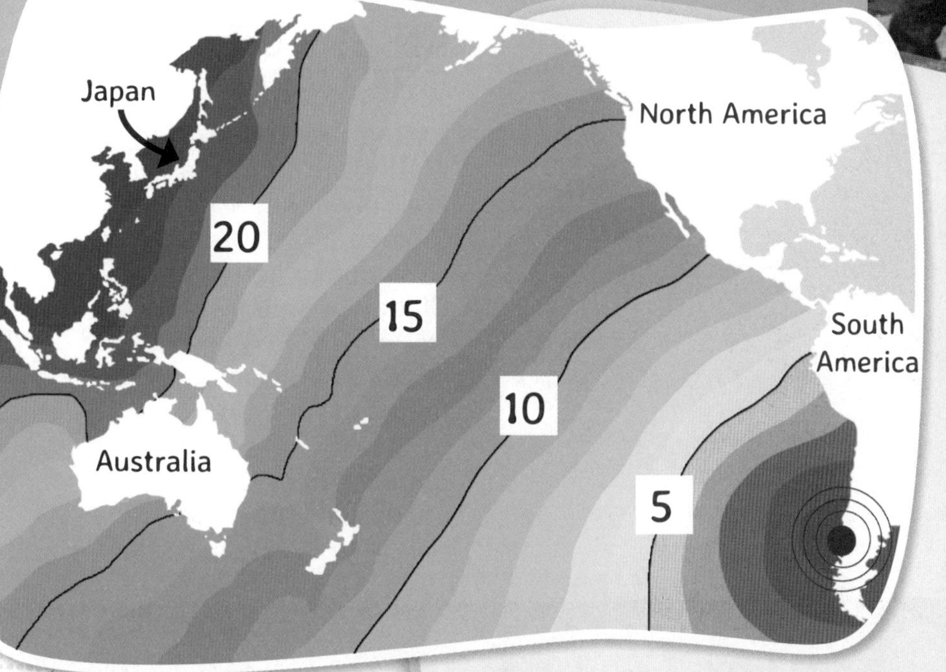

BIG IN JAPAN

Earthquakes as big as Valdivia send their effects worldwide. The tremors from Valdivia could be detected on seismometers on the other side of the world. And the tsunami unleashed travelled right across the Pacific Ocean to Japan in just 22 hours, causing a devastating 5.5-m wave.

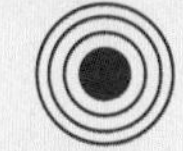 Epicentre

Travel time (in hours) of wave front

DOWN UNDER

Many earthquakes start when tectonic plates slide sideways past each other. The Valdivia quake was different. It started under the sea in a deep ocean trench off the Chilean coast. Here, one plate is 'subducted' or thrust underneath the neighbouring plate. The quake started when the western edge of the South American Plate lurched 18 m up over the Nazca Plate near the base of the trench.

Subducted area ruptures, releasing energy in an earthquake

ONE STEP AHEAD OF THE EARTHQUAKE

Many large cities – such as Los Angeles, Mexico City and Tokyo – are located in areas prone to earthquakes. Sooner or later, one will be hit by a 'Big One'. So the more warning people get, the better. Scientists know why quakes happen, but it is hard to predict when they will.

UNBELIEVABLE!

Do animals know quakes are coming? In 1975, scientists in Yingkou, Liaoning Province in China noticed snakes and rats emerging from their holes. The city was evacuated just before a huge quake struck, and many lives were saved.

AN EYE ON THE WORLD

Earthquake monitoring stations around the world are now linked together to keep a constant eye on earthquake activity. The Global Seismographic Network (GSN) keeps over 150 stations linked via the internet to quickly detect and pinpoint every significant earthquake anywhere.

TESTING THE GROUND

Many seismologists believe the answer is to watch for signs of strain building up in the rocks. In many earthquake zones, high precision surveys now monitor the ground for any signs of deformation in the rocks. Accurate surveys on the surface, for instance, may pick up slight horizontal movements, while tiltmeters set underground may show any vertical shift.

EARLY WARNING

The waves that do most damage in a quake are called S-waves. But waves called P-waves travel much faster and arrive first. In Japan, if a network of monitoring stations detects the P-waves from an approaching quake, they instantly broadcast an alert on TV and radio, and alarms sound in schools and factories.

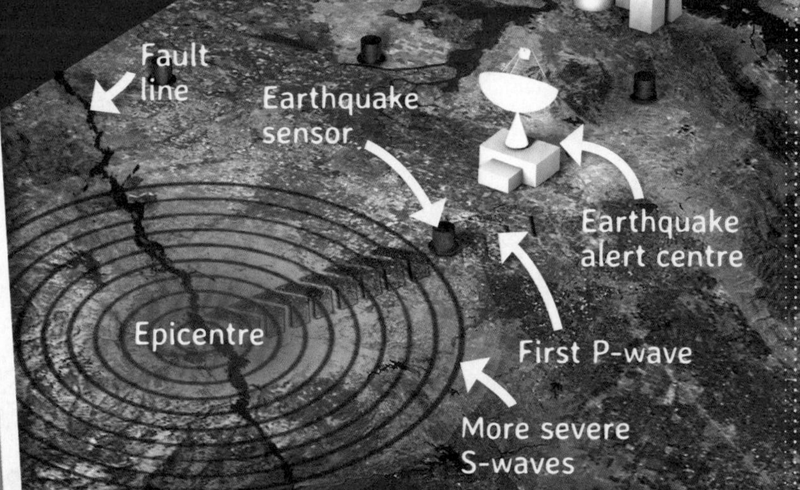

THE POWER OF AN EARTHQUAKE

The Richter scale, devised in 1935 by earthquake scientist Charles Richter, measures the magnitude (size) of an earthquake. Each step from 1 to 10 on the scale indicates a tenfold increase. The typical effects given for each step may vary widely.

Scale	Damage
0–1.9	Can't be felt by people
2–2.9	Felt by some people, no damage
3–3.9	Visible shaking, occasional damage
4–4.9	Rattling, things fall off shelves
5–5.9	Felt by all, damage to rickety buildings
6–6.9	Violent shaking near epicentre, damage to buildings
7–7.9	Damage or collapse of most buildings, felt 250 km away
8–8.9	Most buildings destroyed near epicentre
9 and over	Total destruction over a large area

HOW DOES A TSUNAMI HAPPEN?

Tsunamis are a brief series of waves like the ripples from a stone thrown in a pond – only much, much bigger! A tsunami typically begins with an earthquake – maybe when a big chunk of the Earth's crust snaps in the seafloor, sending pulses of water through the ocean in all directions.

UNBELIEVABLE!

Tsunamis can also be started by the impact of a meteor landing, although this is rare. But there are signs of huge tsunami damage around the Pacific caused by a meteor crashing into the ocean 2.2 million years ago.

HIDDEN TERROR

Out at sea, tsunamis travel unseen along the seabed. But they move faster than a jet airliner. Once they reach shallow water, they may slow down a little, but like water slopping at the edge of a bath, they can rear up to frightening heights.

JAPAN

One or two major tsunamis strike somewhere in the world every year. But Japan gets more than most (Tōhoku, March 2011, shown). Tsunami is Japanese for 'harbour wave', so-called because fishermen never saw these waves at sea — then returned home to find their harbour devastated.

Near the coast, the backwards roll of the tsunami pulls water away from the shore

An earthquake sends water rolling across the seafloor at high speed

EYEWITNESS

The first sign of a tsunami may be the sea suddenly receding. When the Boxing Day 2004 tsunami hit the beach at Phuket in Thailand, 10-year-old British tourist Tilly Smith remembered this from a geography lesson and warned people on the beach to run inland in time to save their lives.

The tsunami rears up and rolls right over, crashing on to the shore

TSUNAMI SPEEDS

The deeper the water in which a tsunami travels, the faster it can move. A deep tsunami can travel right across the Pacific Ocean in less than 20 hours!

Depth (m)	**Speed** (km/h)
7,000	943
4,000	713
2,000	313
200	159
50	79
10	36

VOLCANIC TSUNAMIS

Not all tsunamis are generated by earthquakes. Some are set off by erupting volcanoes. One of the worst tsunamis ever occurred on 26 August 1883, after the explosion and collapse of the volcano of Krakatoa in Indonesia. It sent out waves reaching 40 m, and destroyed coastal towns and villages on Java and Sumatra, killing 36,417 people.

1. Original summit of volcano
2. Volcano collapses
3. Magma body is unroofed
4. Lateral blast
5. Debris crashes into the sea
6. Tsunami forms

EXTREME TSUNAMI DAMAGE

Perhaps no natural disaster is quite so shattering and sudden as a tsunami. It comes suddenly out of the sea with little or no warning to swamp coastlines with giant walls of water that can wash away everything and sweep far inland.

UNBELIEVABLE!

One man survived the 1883 Krakatoa tsunami by jumping on the back of a giant crocodile. He held on to it as the wave swept them inland for 3.2 km and crashed on a hill. He then jumped off and ran for his life.

SWEPT TO SHORE

An indication of the amazing power of a tsunami is shown by this engraving made after a tsunami hit Sumatra in Indonesia following the eruption of Krakatoa in 1883. This steamship was travelling calmly out at sea. Minutes later it had been swept 1.6 km inland. Its crew of 28 all died.

EYEWITNESS

As the Tōhoku tsunami (right) rolled in, people saw their homes being swept away. One witness heard the cry "Tsunami coming!" almost 800 m inland. "I rushed up the stairs to the rooftop of the building with the residents. The tsunami hit the centre. The building was surrounded by water and wreckage of houses and buildings. The water reached as high as the ceiling of the second floor."

TŌHOKU DISASTER

On 11 March 2011, a gigantic earthquake shook the ocean floor 70 km off the east coast of Japan. It set off a tsunami hitting the Sendai region with waves up to 40.5 m tall and roared 10 km inland. A million buildings were destroyed or badly damaged and nearly 16,000 people lost their lives.

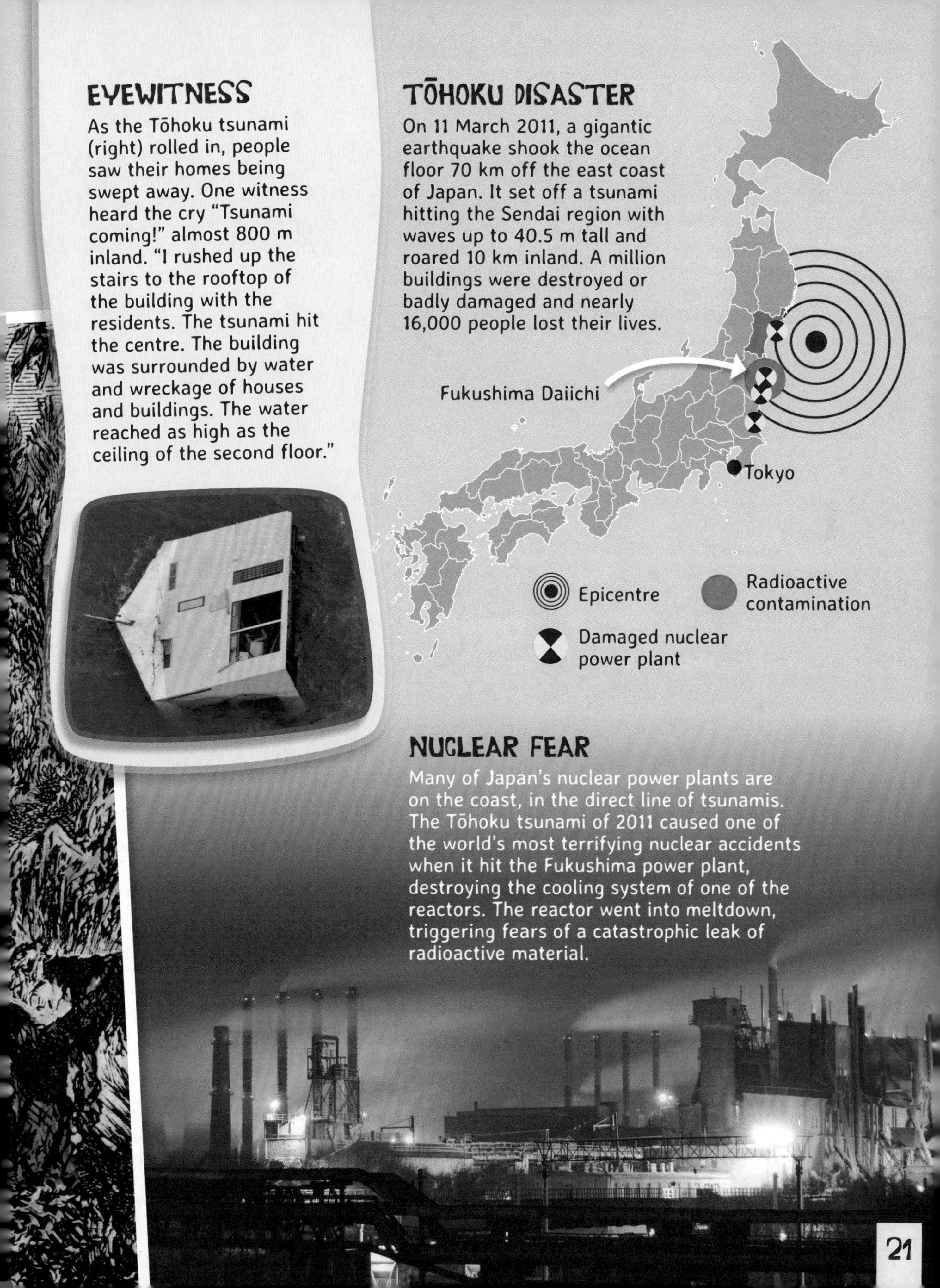

NUCLEAR FEAR

Many of Japan's nuclear power plants are on the coast, in the direct line of tsunamis. The Tōhoku tsunami of 2011 caused one of the world's most terrifying nuclear accidents when it hit the Fukushima power plant, destroying the cooling system of one of the reactors. The reactor went into meltdown, triggering fears of a catastrophic leak of radioactive material.

EXTREME TSUNAMI STORY

On 26 December 2004, the seafloor off the coast of Sumatra in South-east Asia was rocked by one of the biggest earthquake ever recorded. The earthquake itself did little damage on land, but it ripped apart the seabed and lifted it 15 m along a huge stretch of fault in the Sunda Trench, sending a devastating tsunami across the Indian Ocean.

UNBELIEVABLE!

The earthquake that triggered the Boxing Day tsunami was the longest ever – lasting up to 10 minutes. It made the whole Earth wobble up to 2 cm off its axis.

BANDA ACEH

On the shores directly facing the quake, the tsunami reared up to over 30 m and roared over 1.6 km inland. The nearby Sumatran city of Banda Aceh (main picture) was hit by the full force of the tsunami 5 minutes after the quake began, and destroyed almost totally in 15 minutes, with tens of thousands people killed.

STRANDED

This boat left perched on top of two houses gives an idea of just how extreme the effect of the tsunami was – and also the problems with clearing up. Just how do you get a boat off the roof and back to the sea?

DISASTER RELIEF

The Boxing Day tsunami was one of the worst natural disasters ever. Besides claiming the lives of about 230,000 people, it left millions homeless. A massive international programme was launched to rescue and aid survivors.

Bangladesh
India
18,045 dead
Myanmar
400–600 dead
Andaman Islands
Somalia
289 dead
Maldives
108 dead
Sri Lanka
35,322 dead
Thailand
8,212 dead
Malaysia
75 dead
Seychelles
3 dead
INDIAN OCEAN
Indonesia
167,799 dead
Madagascar

OCEAN WIDE

Within half an hour, the tsunami had swamped the Andaman Islands hundreds of kilometres away, and within an hour and half it had overwhelmed the coastal resorts of Thailand with a wall of water 10 m high. In two hours, the tsunami raced thousands of kilometres across the Indian Ocean to hit Sri Lanka and Southern India. Over 100,000 people died in Indonesia, close to the origin, but 289 people died even in Somalia, East Africa.

EMERGENCY AID

After the tsunami, emergency crews rushed into action dropping water, food and medical supplies by helicopter, since many roads were washed away. But the disaster was so widespread, that many people, especially those in remote villages, received no help at all.

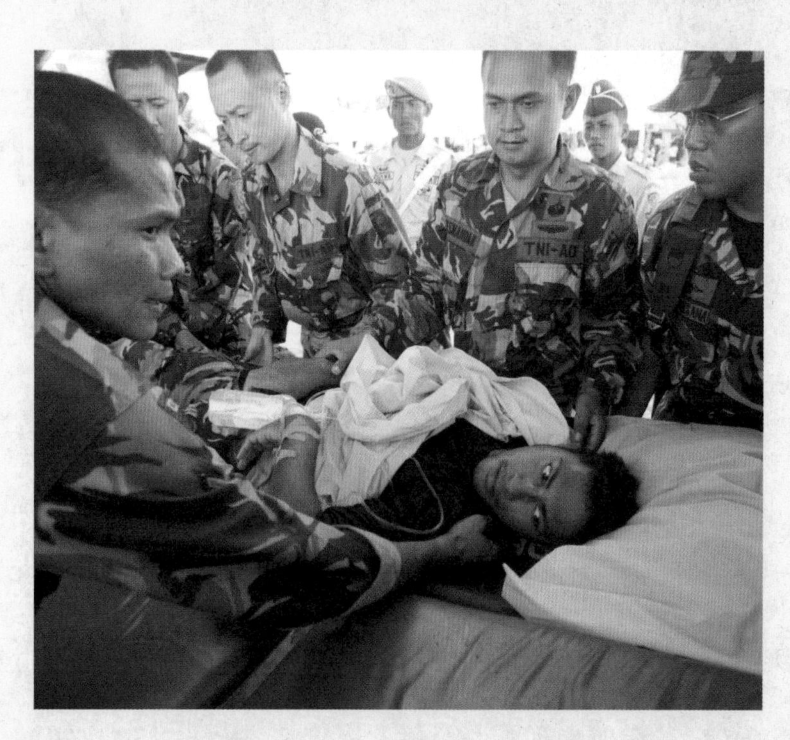

ONE STEP AHEAD OF THE TSUNAMI

Tsunamis can take several hours to travel across the ocean. So if scientists can detect a tsunami early on, they may be able to issue warnings and give people a chance to escape. Nowadays, the Pacific Tsunami Warning System goes on to alert every time the Global Seismographic Network (see p.16) detects a large, shallow quake under the Pacific Ocean.

UNBELIEVABLE!

After the Boxing Day tsunami of 2004, researchers found that much less damage was done where the shoreline was protected by coral reefs or vegetation. Human activity weakens these natural defences through pollution and deforestation, for example.

TSUNAMI BUOYS

The DART system is a series of 40 or so buoys dotted across the ocean and linked by sound waves to special pressure sensors on the ocean floor. These sensors can detect a slight rise or fall in the depth of water above — even as little as 1 mm. If there is a sudden change, the buoy sends out the alarm via a satellite link.

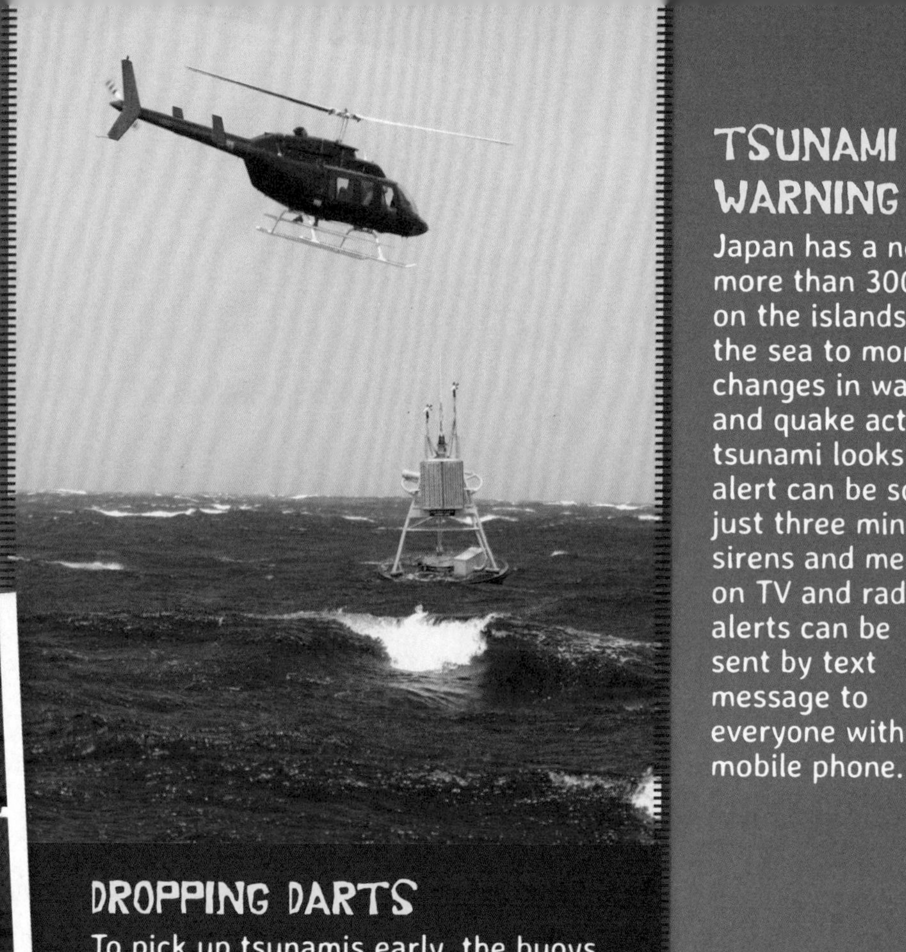

TSUNAMI WARNING

Japan has a network of more than 300 sensors on the islands and in the sea to monitor changes in water level and quake activity. If a tsunami looks likely, an alert can be sounded in just three minutes with sirens and messages on TV and radio. Now alerts can be sent by text message to everyone with a mobile phone.

DROPPING DARTS

To pick up tsunamis early, the buoys have to be near the places where they start. That means they must be scattered across the ocean floor — particularly near deep trenches in the ocean floor where earthquakes often happen. They are dropped at the right location by helicopters and anchored in place by weighted cables.

SEA DEFENCE

Early warning may help people to evacuate coastal areas before a tsunami strikes, but the impact of tsunamis can also be reduced by natural and artificial barriers on the shore. In the city of Numazu in Japan, they have built a giant gate across the dock entrance. It can be dropped closed in just five minutes to block off a tsunami as much as 6 m high.

INTENSE FUTURES

Scientists have put a lot of effort into trying to work out when and where the next big earthquake or tsunami will strike. The best they can do so far is give us a little warning once it has already begun.

MONSTER WAVE

Cumbre Vieja is a volcano on the Canary Islands in the Atlantic. One day in the future some of it might suddenly collapse into the sea. If so, could it unleash a mega-tsunami which could sweep across the Atlantic? Vast waves hundreds of metres high might crash over American cities of Boston, New York and Miami as in this imaginary picture. Luckily, most scientists think this is unlikely.

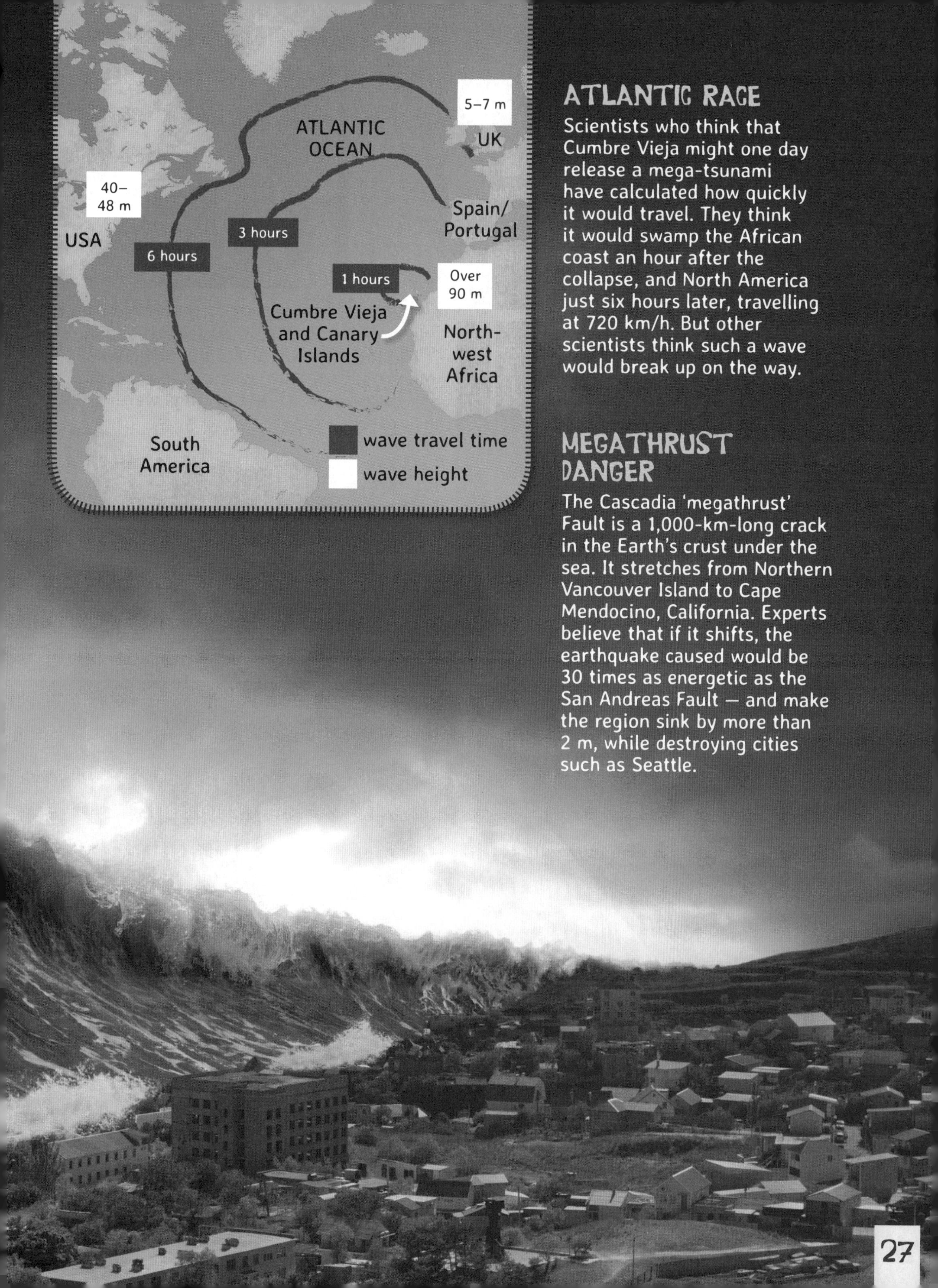

ATLANTIC RACE

Scientists who think that Cumbre Vieja might one day release a mega-tsunami have calculated how quickly it would travel. They think it would swamp the African coast an hour after the collapse, and North America just six hours later, travelling at 720 km/h. But other scientists think such a wave would break up on the way.

MEGATHRUST DANGER

The Cascadia 'megathrust' Fault is a 1,000-km-long crack in the Earth's crust under the sea. It stretches from Northern Vancouver Island to Cape Mendocino, California. Experts believe that if it shifts, the earthquake caused would be 30 times as energetic as the San Andreas Fault – and make the region sink by more than 2 m, while destroying cities such as Seattle.

TIMELINE

1906
After the Great San Francisco Earthquake, a firestorm raged and left many homeless

132 AD
Zhang Heng invented the first earthquake detector (replica shown)

1923
Much of Tokyo, Japan, was destroyed in a fire after the Great Kanto Earthquake. More than 140,000 people died or went missing and 360,000 buildings were destroyed (below)

132 AD

1775
After an earthquake in Portugal, a huge tsunami hit its capital Lisbon, with waves over 15 m high

1556
A massive earthquake hit Shaanxi and Shanxi provinces in China, killing 830,000 people

1885
British geologist John Milne invented the first modern seismograph

1935
American Charles Richter invented the Richter scale to measure earthquake magnitude

1960
The largest earthquake ever recorded, 9.4–9.6 on the Richter scale, hit Chile, creating tsunamis that caused damage as far away as Japan

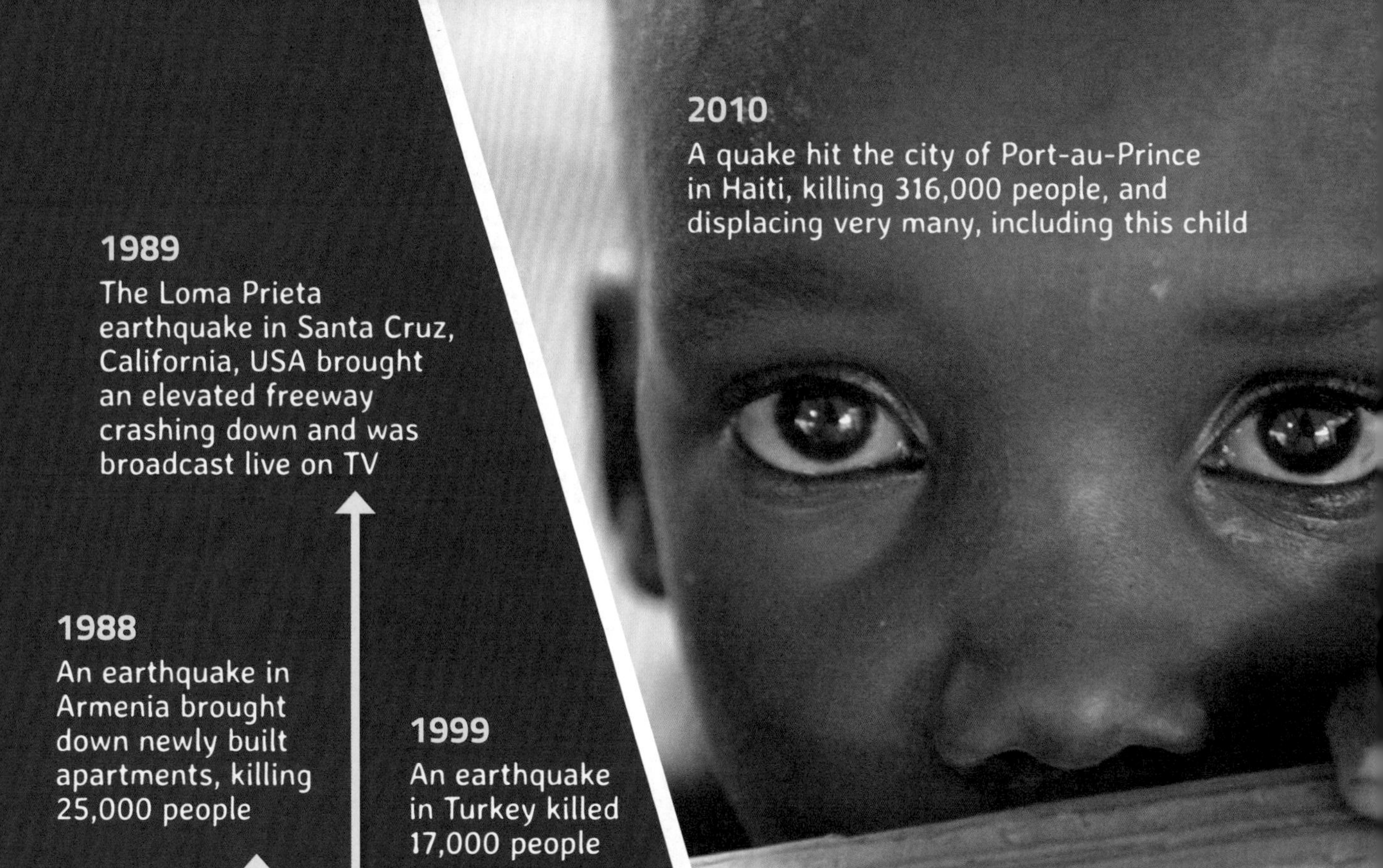

2010
A quake hit the city of Port-au-Prince in Haiti, killing 316,000 people, and displacing very many, including this child

1989
The Loma Prieta earthquake in Santa Cruz, California, USA brought an elevated freeway crashing down and was broadcast live on TV

1988
An earthquake in Armenia brought down newly built apartments, killing 25,000 people

1999
An earthquake in Turkey killed 17,000 people

1970
An earthquake in Peru set off a massive avalanche in the mountains that killed about 70,000 people

1976
An earthquake measuring 7.8 hit Tangshan, China, killing up to 240,000 people

1995
In Kōbe, Japan, 6,400 people died in a 7.2 earthquake

2001
An earthquake in Gujarat, India, killed 20,000 people

2003
An earthquake in Iran killed 26,000 people and destroyed the ancient city of Bam

2004
A huge earthquake off Sumatra, Indonesia, triggered the Boxing Day tsunami that killed about 230,000 people around the Indian Ocean

2005
A 7.6 earthquake in Pakistan and Kashmir killed over 79,000 people

2011
The Tōhoku earthquake off the east coast of Japan caused terrible tsunamis

BLOWN AWAY!

Amazing facts about earthquakes and tsunamis

LOST TIME

Earthquakes can make the whole Earth move! The 2011 Tōhoku quake in Japan shifted the Earth's mass and made the Earth spin faster, shortening the day by 1.6 microseconds. The 2004 Sumatra quake shortened the day by 6.8 microseconds.

BRIGHT SPARKS

Three Greek scientists, professors Varotsos, Alexopoulos and Nomikos, developed a system to predict earthquakes known as the VAN method after their initials. It looks for disturbances in natural electric currents, called telluric currents, flowing in the ground. They managed to predict some quakes this way. But most scientists are not yet convinced by the idea.

MOONQUAKE

The moon has quakes, too – only they are called moonquakes, not earthquakes. They are normally weaker than earthquakes.

A TINY SHIFT

Tectonic plates, those giant slabs of rock that make up the Earth's surface, move very slowly – less than 17 cm per year. But a tectonic plate only has to budge 20 cm or so to set off a major earthquake.

NOSE FOR A QUAKE

Sometimes, ponds and canals give off a strange smell before an earthquake. Scientist think the smell comes from the release of gases underground. The water on the ground can also become warmer.

Mount Bromo, Indonesia, sits on the Ring of Fire

QUAKES IN 3D

In recent years, earthquake scientists have begun to bring the massive calculating power of supercomputers to bear on predicting earthquakes. The Quake Project at the Southern California Earthquake Centre in the USA is working with one of the world's fastest computers to build a 3D computer model of just what goes on in the ground during an earthquake. It uses very detailed seismograph data.

EARTHQUAKE HOTSPOTS

Four of every five big quakes happen around the 'Ring of Fire'. This is a huge ring around the Pacific Ocean where many tectonic plates meet. The second most earthquake-prone area is the Alpide Belt, which spans Turkey, India and Pakistan.

WORST EARTHQUAKES

WORLD'S DEADLIEST

Shaanxi and Shanxi provinces, China
23 January 1556
Cost in lives: over 830,000

DEADLIEST RECENT

Tangshan, China
28 July 1976
Cost in lives: up to 240,000

MOST POWERFUL

Valdivia, Chile
22 May 1960
Magnitude: 9.4–9.6 on the Richter magnitude scale

MOST POWERFUL IN THE USA

Prince William Sound,
south-central Alaska
27 March 1964
Magnitude: 9.2 on the Richter magnitude scale

WORST TSUNAMIS

WORLD'S DEADLIEST

Crete earthquake tsunami,
Eastern Mediterranean, Greece
21 July 365 AD
Cost in lives: maybe up to 500,000

DEADLIEST RECENT

Boxing Day tsunami, Indian Ocean
26 December 2004
Cost in lives: 230,000

TALLEST

Lituya Bay tsunami, USA
9 July 1958
Height: 524 m

FASTEST

Boxing Day tsunami, Indian Ocean
26 December 2004
Speed: 800 km/h

INDEX

THE AUTHOR

John Farndon is Royal Literary Fellow at City&Guilds in London, UK, and the author of a huge number of books for adults and children on science, technology and nature, including such international best-sellers as *Do You Think You're Clever?* and *Do Not Open.* He has been shortlisted six times for the Royal Society's Young People's Book Prize for a science book, with titles such as *How the Earth Works, What Happens When...?* and *Project Body* (2016).

Picture Credits (abbreviations: t = top; b = bottom; c = centre; l = left; r = right)
© www.shutterstock.com:
1 c, 4 c, 14 b, 19 br, 21 br, 26-27 c, 28 tr, 28 cr, 30 bl, 32 r.
11 tr = © Antonio Nardelli / Shutterstock.com, 13 tr - © Fotos593 / Shutterstock.com, 13 br = © fabiodevilla / Shutterstock.com, 29 tr = © arindambanerjee / Shutterstock, Inc, 31 tl = © CHEN WS / Shutterstock, Inc.

BC, cr = Design Pics Inc / Alamy Stock Photo. 2, l = epa european pressphoto agency b.v. / Alamy Stock Photo. 3, c = Pictura Collectus / Alamy Stock Photo. 6, c = Chronicle / Alamy Stock Photo. 7, c = Design Pics Inc / Alamy Stock Photo. 8, b = Kevin Schafer / Alamy Stock Photo. 9, cl = epa european pressphoto agency b.v. / Alamy Stock Photo. 10, c = Christian Kober 1 / Alamy Stock Photo. 11, cr = REUTERS / Alamy Stock Photo. 12, c = Bart Pro / Alamy Stock Photo. 13, b = michael seaman / Alamy Stock Photo. 13, cl = Shutterstock.com / deepspace. 15, tr = ZUMA Press, Inc. / Alamy Stock Photo. 16, c = dpa picture alliance archive / Alamy Stock Photo. 17, tl = Chrispo / Alamy Stock Photo. 18, c = Archive Image / Alamy Stock Photo. 20, c = North Wind Picture Archives / Alamy Stock Photo. 21, tl = Pictura Collectus / Alamy Stock Photo. 22, c = epa european pressphoto agency b.v. / Alamy Stock Photo. 22, b = REUTERS / Alamy Stock Photo. 23, cl = Design Pics Inc / Alamy Stock Photo. 23, br = epa european pressphoto agency b.v. / Alamy Stock Photo. 25, br = Photo Japan / Alamy Stock Photo. 25, tl = david hancock / Alamy Stock Photo. 25, tr = imageBROKER / Alamy Stock Photo. 28, bl = Granger Historical Picture Archive / Alamy Stock Ph. 28, tl = Keren Su/China Span / Alamy Stock Photo

番茄博士双语百科

极端自然灾害

地震与海啸

约翰·法恩登（英） 著
陈泰之 译

外语教学与研究出版社
FOREIGN LANGUAGE TEACHING AND RESEARCH PRESS
北京 BEIJING

地震能将整座城市移动3米，能让地球转得更快，使世界为之颤抖。这是原始的力量在发力，随之而来的是惊涛骇浪……它们能够行进上千千米，跨越大洋，将船只抛至陆地，刹那间毁灭上百万座房屋。

了解一下最可怕的地震和海啸，它们的成因和破坏性影响，以及探测这些隐而不发的恐怖事物的最新技术。实际上这个星球上的每个地方都有可能发生地震或海啸。

目录

地震

地震是地表的剧烈震动，构成地球表面的巨大岩石块发生突然断裂或移动，引起了地震。多数地震非常微弱，只有最灵敏的设备才能监测到。但也有一些地震的威力非常强大，它们能够造成破坏，摧毁城市，夺去许多人的生命。

海啸

海底发生地震或发生火山喷发时，能够引起大规模海水运动，这些运动产生的大浪或一系列海浪称为海啸。海水向四面八方涌动，我们看不到这些发生在海底的水流运动。但是，一旦这些海浪接近海岸，就会形成巨大的带有破坏性的水墙，冲走行进道路上的一切东西。

地震是怎样发生的？

地球的固体表面（也叫“地壳”）断裂形成若干块巨大的岩石，称作地壳板块。这些板块之间互相摩擦。大多数地震发生在板块突然破裂或错动时。此时，它们会向地面释放出冲击波或“地震波”。

你知道吗？

没有人能跑得过地震波。地震波沿地表的传播速度最快可达每秒8千米——相当于喷气式飞机速度的40多倍！

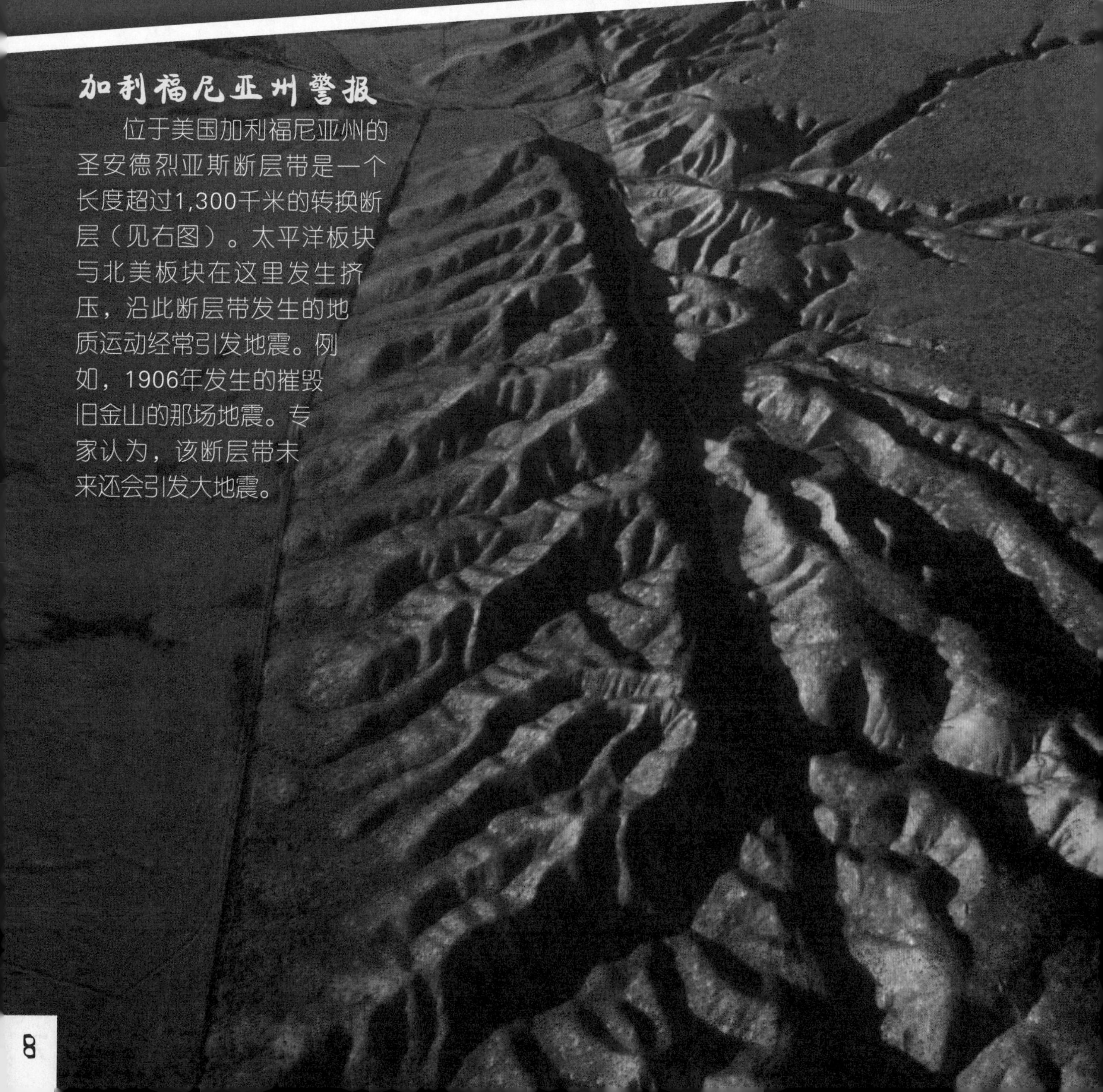

加利福尼亚州警报

位于美国加利福尼亚州的圣安德烈亚斯断层带是一个长度超过1,300千米的转换断层（见右图）。太平洋板块与北美板块在这里发生挤压，沿此断层带发生的地质运动经常引发地震。例如，1906年发生的摧毁旧金山的那场地震。专家认为，该断层带未来还会引发大地震。

震动的板块

一些板块运动时会相互推开，一些会挤在一起，一些会侧滑而过。正是这些称为“转换断层”的滑动连接部位引发了大部分地震。锯齿状的板块边缘互相挤在一起，压力不断增大，直至板块接壤处突然发生滑脱，引发地震。

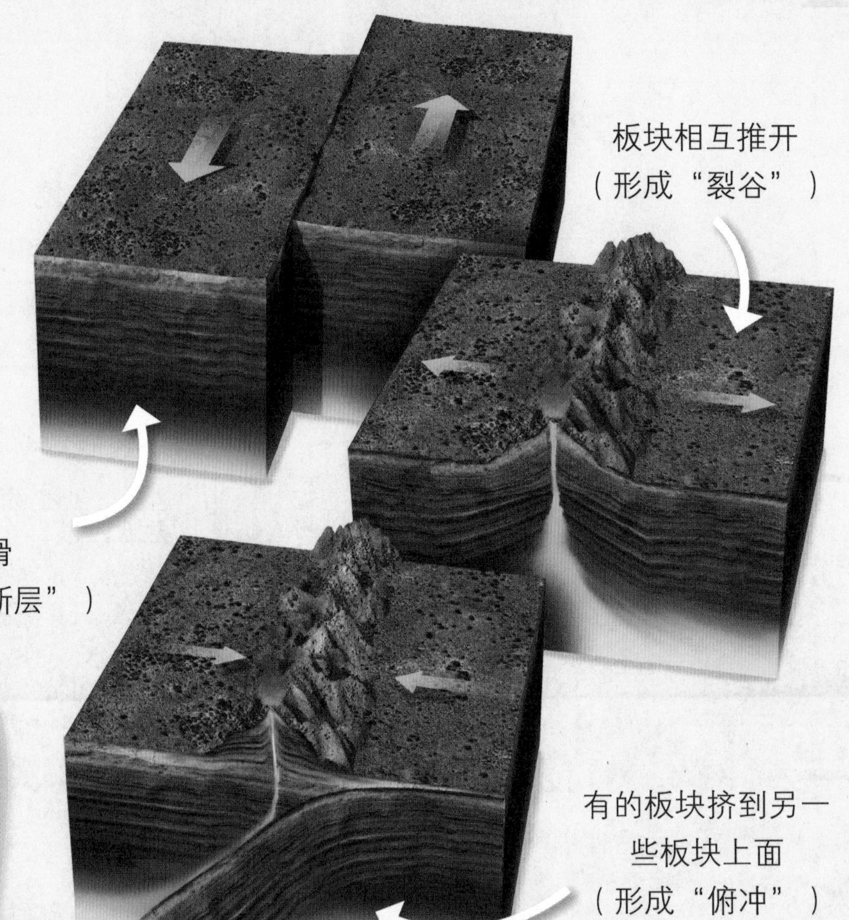

海底火山

大多数地震是由板块之间的运动引发的，也有一些地震是由火山喷发导致的。炽热的岩浆涌入火山底部的岩缝时，导致岩石突然断裂，从而引发地震。

地震带

世界上没有地方能够完全避开地震的影响，但多数大地震发生在地壳板块边缘的地震带上。大约80%的大地震发生在太平洋的边缘地带，其余的许多大地震发生在欧洲南部、近东地区和亚洲南部。

世界板块构造图

地震的危害

地震发生时，地面剧烈地来回晃动。在浅源大地震的震中位置，震感十分强烈，最坚固的建筑物也难以幸存。地震撕裂地面，使地面左右或上下晃动，甚至崩坍。地震发生时，岩石断裂释放的热量有时能将地表熔化。

你知道吗？

2010年，智利发生的一次地震使整个康塞普西翁市向西移动了3米，而且略微缩短了地球一天的时长！

海地的悲剧

2010年1月12日，地震袭击了岛国海地及其首都太子港（见大图），造成了十分可怕的后果。建筑质量不佳的房屋倒塌，超过316,000人在地震中丧生，3,000,000人的家园被毁。地震使供水中断，导致霍乱流行。

火灾损失

地震中燃气管道和电缆受损，经常引起火灾，造成一些最严重的损坏。1906年4月18日，旧金山遭遇了大地震，这是美国历史上最严重的自然灾害之一。但是，大部分的损失是由震后持续了三天的大火造成的。

突如其来的灾难

地震最可怕的一点是它的发生让人毫无防备。即便是地震多发区也可能数百年不会发生一次地震。这正是意大利的历史名城阿马特里切的遭遇。2016年8月，在没有任何预警的情况下，该城在几分钟内被地震完全摧毁。

断桥

桥梁很容易被地震损毁。一个轻微的地壳运动就能折断大桥，造成毁灭性的后果。因此，地震带的工程师会把桥梁建得尽可能抗震，例如，旧金山海湾大桥。

大地震发生之后

地震的持续时间一般不会超过一分钟，但就在这么短的时间内，它们就能造成可怕的破坏。震后救援工作越早开展越好。但是，由于道路和水电供应中断，以及建筑物可能会发生二次坍塌，救援工作面临着巨大挑战。

你知道吗？

2010年，全球各国人民一共募集了35亿美元，以援助海地地震灾民。有些电影演员每人捐了100万美元，巴西模特吉赛尔·邦辰捐了150万美元，美国高尔夫运动员泰格·伍兹捐了300万美元。

重建

震后重建是一项长期工作，尤其是在资源匮乏的贫穷国家。2010年的海地地震发生两年后，仍有50万人无家可归，霍乱疫情也未得到有效控制。

探测设备

现代科技对搜寻幸存者很有帮助。摄像机可以绑在细杆上，探进狭小的孔洞。热成像仪能在黑暗中探测到人的体温。专业的声音设备能够接收到人的呼吸声，前提是救援人员必须保持安静。

搜救犬

震后救援的首要任务之一是营救困在倒塌房屋下面的幸存者。为了尽快找到他们，救援人员可能会使用嗅探犬。嗅探犬具有敏锐的嗅觉，能够捕捉到救援人员发现不了的生命迹象。此外，嗅探犬还能快速地搜寻大片区域。

外部救援

灾民需要救援，震区需要重建。地震发生后需要立即向震区派遣赈灾人员、搜救团队、医务人员、技术人员和安保人员。世界各地民众可以筹集资金，为灾民购买饮用水、新家具等一切所需物资。

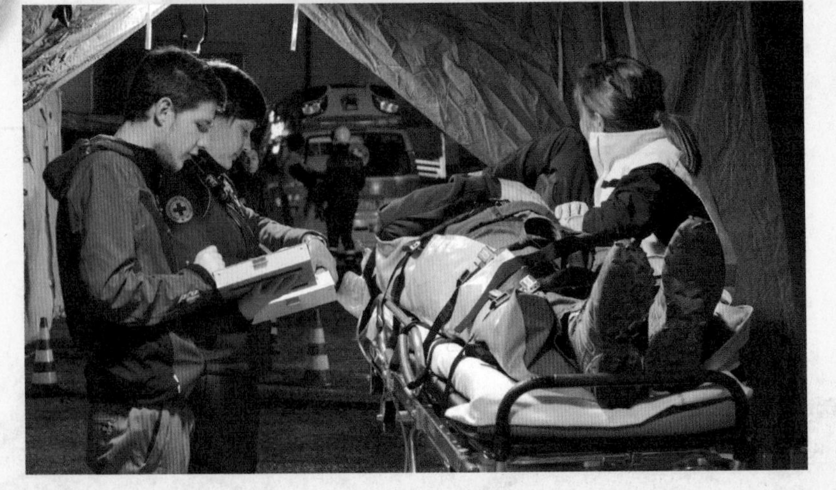

医疗援助

地震发生之后，医护人员至关重要。医护人员不仅能够治疗重伤，还需要应对因断粮断水引发的各种疾病。震后灾区饮用水受到污染，容易引发霍乱、痢疾和其他通过水传播的疾病。

大地震的故事

1960年5月22日，地震袭击了智利的瓦尔迪维亚镇。这是史上记录到的最强烈的一次地震。下午3点11分，地震始发于距离海岸160千米处的海洋中，瞬间波及瓦尔迪维亚镇及其他沿岸城镇。但是，沿岸地区的主要损失是由地震发生15分钟后袭来的大海啸造成的。

你知道吗？

1960年智利大地震释放的能量，相当于20世纪发生的全部地震以及2004年印度洋大地震所释放的所有能量的四分之一。

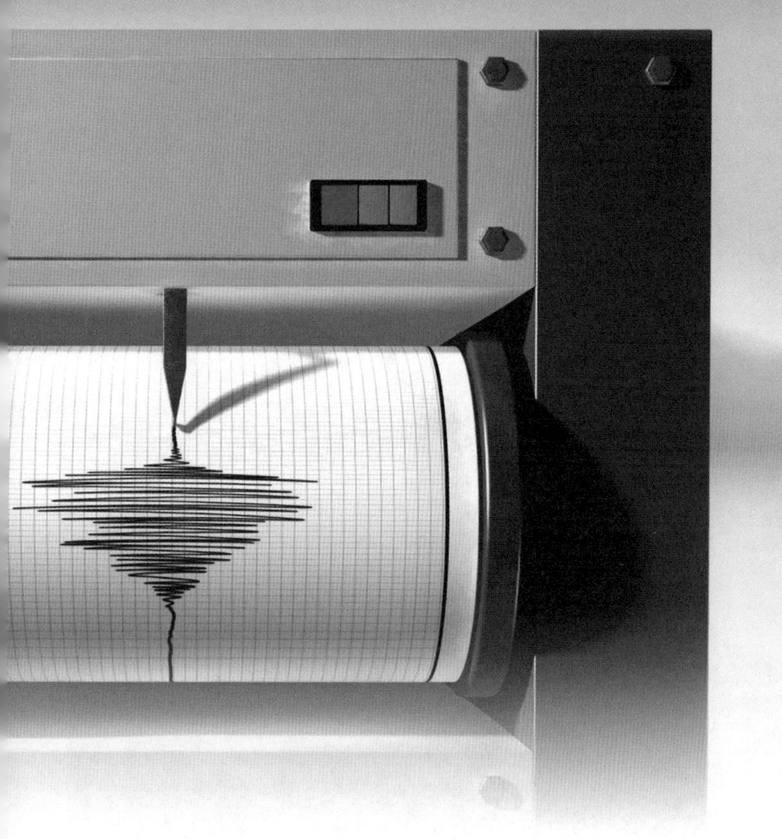

地震的大小

科学家使用“地震计”记录地震的震动或地震波。过去，他们用里氏震级标记地震等级。里氏震级包括从0到9甚至更高，数字越大，地震等级越高。现在，对于最大的地震，他们使用矩震级来标记。矩震级将里氏震级与岩石运动结合起来，能够反映地震的真正威力。

强震

瓦尔迪维亚地震的矩震级是9.4—9.6级。除此次地震之外，只有六次地震的矩震级达到了9级。

1960年5月22日	智利，瓦尔迪维亚	9.4—9.6级
1964年3月27日	美国，阿拉斯加州威廉王子湾	9.2级
2004年12月26日	印度尼西亚，苏门答腊	9.1—9.3级
2011年3月11日	日本，东北部	9.1[4] 级
1952年11月4日	苏联，堪察加	9.0级
1868年8月13日	智利，阿里卡（当时属于秘鲁）	9.0级
1700年1月26日	太平洋，卡斯凯迪亚俯冲带	8.7—9.2级

1960年瓦尔迪维亚地震

日本
北美洲
20
15
10
5
南美洲
澳大利亚

震中　波浪前锋的行进时间（小时）

在日本的巨大威力

像瓦尔迪维亚这样的大地震会造成世界范围的影响。瓦尔迪维亚地震引发的震动能被地球另一端的地震计监测到。它引发的海啸仅用了22小时就跨越了太平洋，抵达日本，并带去了5.5米高的破坏性海浪。

下冲

许多地震发生在板块互相侧滑错动的时候。瓦尔迪维亚地震却不是这样。它发生在智利海岸以外大洋深处的海沟里。在这里，一个板块冲入了相邻板块的下边。纳斯卡板块在海沟底部突然俯冲至南美洲板块西部边缘的下方，这时地震发生了，导致南美洲板块抬升了18米。

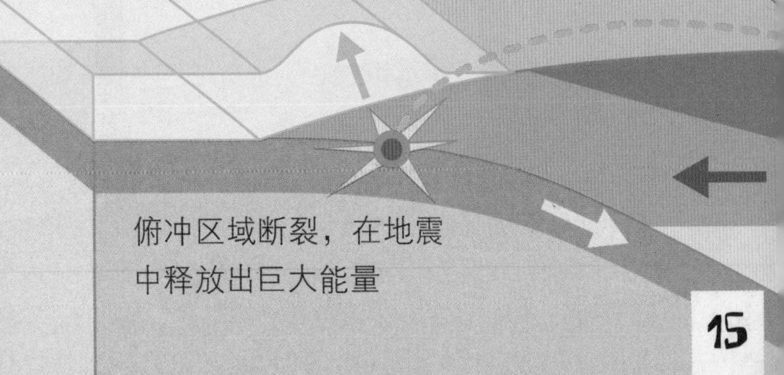

俯冲区域断裂，在地震中释放出巨大能量

地震发生前的预警

许多大城市，如洛杉矶、墨西哥城和东京，都位于地震多发区。迟早会有某个大城市遭遇“大地震”的袭击。因此，人们得到的预警越多越好。科学家知道地震发生的原因，却很难预测它们发生的时间。

你知道吗？

动物能够预知地震的发生吗？1975年，中国辽宁省营口市的科学家们注意到蛇鼠逃离洞穴的现象。该市在一次大地震前疏散了人口，挽救了许多人的生命。

放眼世界

现在，世界各地的地震监测站紧密联系，时刻关注着地震活动。全球地震台网（GSN）通过互联网连接了超过150家地震台，能够快速探测和精确定位世界各地发生的较大等级的地震。

地面监测

许多地震学家认为，地震预警的关键在于观察岩石内部积聚的压力。许多地震带使用高精度的探测仪来监测地面上是否存在岩石的变形迹象。比如，精确地观测地表可以捕捉细微的水平运动，而放置于地下的倾斜仪可以发现任何垂直移动。

预警

地震中破坏力最强的地震波叫横波（S波），但称为纵波（P波）的地震波传播速度更快，所以更早到达。在日本，如果地震监测网络监测到即将发生的地震的纵波，它们会立即在电视台和广播电台发出警报，学校和工厂也会拉响警报。

地震的威力

里氏震级由地震学家查尔斯·里克特在1935年提出，目的是为了衡量地震的震级。从1到10每增加一级，表明地震的强度增加十倍。每一级之间的地震破坏力差别很大。

震级	破坏程度
0—1.9级	无感
2—2.9级	有些人有感，但不造成危害
3—3.9级	明显的震动，有时会造成损害
4—4.9级	物体相互碰撞挤压，咣当作响，架子上的东西会掉落
5—5.9级	普遍有感，不坚固的建筑物受损
6—6.9级	震中附近有剧烈震动，建筑物受损
7—7.9级	大部分建筑物受损或倒塌；震中250千米以外仍有感觉
8—8.9级	震中附近的大部分建筑物被摧毁
9级以上	大片区域被彻底摧毁

海啸是怎样发生的？

海啸是一阵阵波浪，就像石头投进池塘里荡起的涟漪一样——只不过海啸的威力远远超过涟漪！海啸通常始于地震——当地球的一大块地壳在海底断裂时，海洋的水流会涌向四面八方。

你知道吗？

海啸也会因流星坠落地球而形成，尽管这很罕见。有证据表明，220万年前，一颗陨落在太平洋的流星引起了海啸，造成了巨大危害。

看不见的恐怖

在海里，海啸沿着海床潜行。但它们的行进速度比喷气式客机还要快。一旦到达浅水区，它们可能会放慢速度，但就像拍向浴室边缘的水会溅起浪花一样，海啸也会溅起波浪，只不过这些波浪高度惊人。

日本

世界上某些地区平均每年会遭遇一至两次大海啸。而日本遭遇的海啸更多（上图为2011年3月日本东北部海啸的场景）。海啸在日语中为“津波”，意为“港湾中的波浪”。如此称谓是因为渔民在海上并未见到这些波浪，但返家后却发现港湾已经被毁。

在海岸附近，退去的海浪将海水卷离海岸

地震使海底的海水高速流动

目击者

海啸的第一个征兆可能是海水突然退去。2004年，当“节礼日”海啸（海啸发生的日期是12月26日，刚好是节礼日，因此以它命名，译者注）袭击泰国普吉岛海滩时，海滩上一名10岁的英国游客蒂莉·史密斯想起自己在地理课上学过的知识，她及时劝告海滩上的人们往岸上跑，拯救了他们的生命。

海啸使海浪冲天而起，翻滚着撞击海岸

海啸的速度

海啸发生海域的水越深，它的速度就越快。深海海啸可以用不到20个小时的时间穿越太平洋！

水深（单位：米）	速度（单位：千米/小时）
7,000	943
4,000	713
2,000	313
200	159
50	79
10	36

火山海啸

海啸并不都是由地震引起的，有些海啸是由火山喷发引起的。1883年8月26日，在印度尼西亚喀拉喀托火山喷发和坍塌之后，历史上最惨重的海啸之一发生了。海啸中的波浪达到40米高，摧毁了爪哇和苏门答腊的沿海城镇与村庄，造成36,417人死亡。

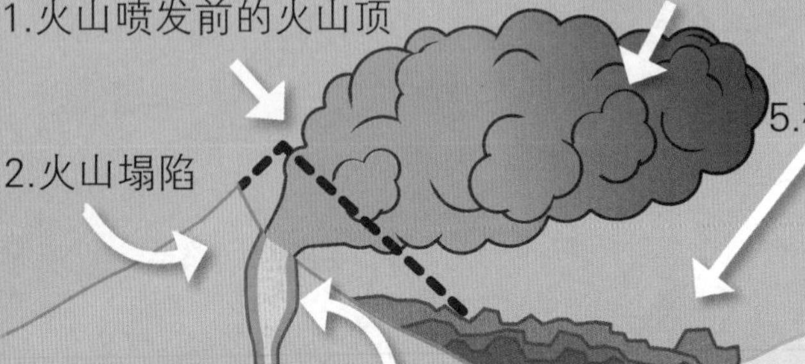

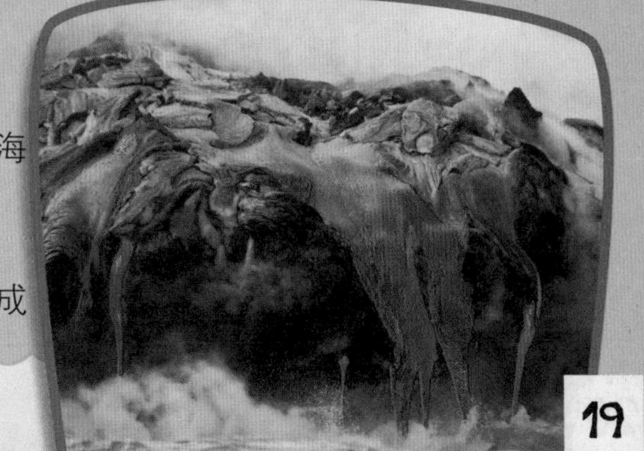

大海啸的危害

海啸或许是最具有毁灭性又最让人猝不及防的自然灾害。它几乎没有或根本没有预警地从海上突然出现，巨大的水墙淹没了海岸线，冲走一切，并涌向内陆深处。

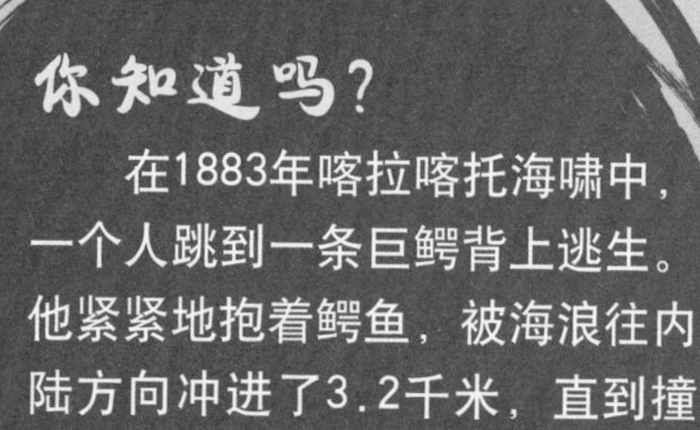

你知道吗？

在1883年喀拉喀托海啸中，一个人跳到一条巨鳄背上逃生。他紧紧地抱着鳄鱼，被海浪往内陆方向冲进了3.2千米，直到撞上一座小山才停下来。他从鳄鱼身上跳下来，逃走了。

卷上岸边

1883年的喀拉喀托火山喷发之后，海啸袭击了印度尼西亚的苏门答腊。此后不久制作的这幅木刻画（见大图）体现了海啸的巨大威力。这艘本来平静地在海上行驶的汽船，几分钟之内被冲到距离海岸线约1.6千米的陆地上，船上的28名船员全部遇难。

目击者

当日本东北部大海啸（见右图）袭来时，人们眼看着自己的房屋被冲走。一位目击者在离岸边不到800米的地方听到“海啸来了！”的叫喊声。“我和其他住户一起爬上楼梯，冲到楼顶。海啸冲击着楼房，楼房被海水包围，水里夹杂着房屋、建筑物的碎片。水面升到了二层楼楼顶的高度。”

日本东北部大灾难

2011年3月11日，一场大地震撼动了日本东海岸以外70千米处的海底。地震引发的海啸袭击了仙台地区，海啸带来的海浪高达40.5米，往陆地方向推进了10千米。上百万座建筑物被毁或严重受损，将近16,000人在地震中丧生。

核恐惧

日本的许多核电站位于沿海地区，在海啸的必经之路上。2011年日本东北部海啸引发了世界上最可怕的核事故之一。海啸袭击了福岛核电站，摧毁了其中一个反应堆的冷却系统。反应堆熔毁，放射物泄漏可能会带来灾难性的后果，这引发了人们的担忧。

大海啸的故事

2004年12月26日，东南亚苏门答腊岛海岸外的海底发生了史上最大的地震之一。地震本身对陆地的破坏不大，但它撕裂海床，把海床沿着巽他海沟处的大片断层提高了15米，制造了一个跨印度洋的灾难性大海啸。

你知道吗？

引发“节礼日”海啸的地震是历史上持续时间最长的一次地震，全程长达10分钟。它使整个地球偏离自转轴线2厘米。

班达亚齐

在正对震中的海岸上，海啸高达30多米，向内陆地带推进了1,600多米。附近的苏门答腊岛上的城市班达亚齐（见大图），在地震发生5分钟后遭遇了海啸的全力打击。15分钟之后该地几乎被全部摧毁，数万人丧生。

搁浅

这条船停在了两座房屋的屋顶上，让人着实感受到海啸的威力，也直观说明了灾后清扫的问题。比如，怎么才能把船从屋顶上弄回到海里？

赈灾

“节礼日”海啸是史上最严重的自然灾害之一。它不仅夺去了大约230,000人的生命，还使数百万人无家可归。为救助受灾人员，人们发起了一项大规模的国际救助计划。

孟加拉国

印度
18,045人丧生

缅甸
400—600人丧生

安达曼群岛

索马里
289人丧生

马尔代夫
108人丧生

斯里兰卡
35,322人丧生

泰国
8,212人丧生

马来西亚
75人丧生

塞舌尔
3人丧生

印度尼西亚
167,799人丧生

印度洋

马达加斯加

跨越大洋

不到半个小时，海啸就已经淹没了数百千米以外的安达曼群岛。一个半小时之内，10米高的水墙冲击了泰国的海滨旅游胜地。两小时后，海啸在印度洋上穿越数千千米，抵达斯里兰卡和印度南部。离海啸发源地较近的印度尼西亚有100,000以上的人口丧生。即便是在东非的索马里，仍有289人失去了生命。

紧急救助

海啸过后，由于许多道路被冲毁，救援人员迅速采取行动，用直升机投送水、食物和医疗用品。但是，这场灾难波及范围巨大，许多人，尤其是那些住在偏远的村庄的人，无法得到任何帮助。

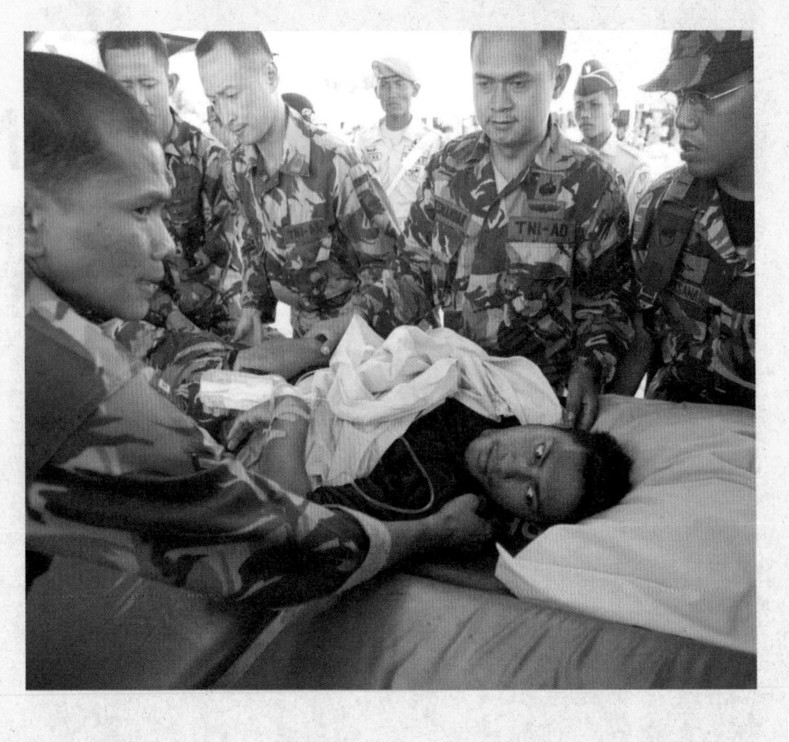

海啸发生前的预警

海啸穿越海洋需要几个小时的时间。因此，如果科学家能够及早发现海啸，也许就能发出警报，给人们逃生的机会。如今，一旦全球地震台网（见第16页）探测到太平洋中大的浅源地震，太平洋海啸预警系统就会发出警报。

你知道吗？

2004年的“节礼日”海啸过后，研究人员发现，有珊瑚礁或植物保护的海岸线遭到的破坏要小得多。污染和森林砍伐等人类活动削弱了这些天然屏障的保护能力。

海啸浮标

“深洋海啸监测与预警系统”由一系列的40个左右的、散布在大洋上的浮标构成。这些浮标通过声波与置放于大洋底部的特制压力传感器相连。这些传感器甚至可以检测到小至1毫米的水位变化。如果水位发生突然的变化，浮标就会通过卫星发出警报。

海啸预警

日本建有由陆地上和海洋里的300多个传感器构成的网络，以监测水位和地震活动的变化。如果有发生海啸的可能，网络系统3分钟内就会发出警报，电视和广播同步发布信息。现在，警报能以短信形式发给每一个有手机的人。

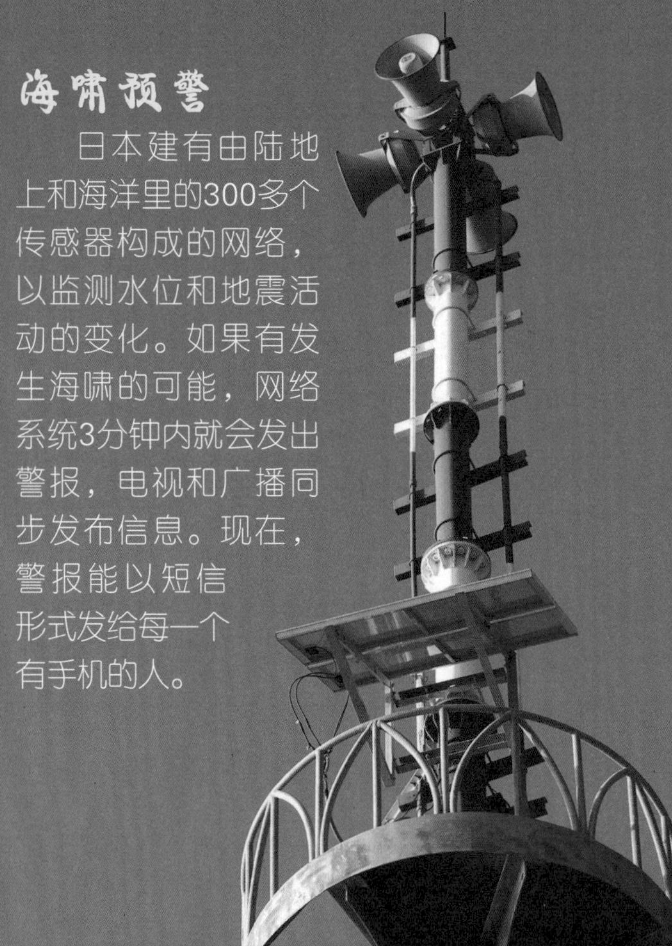

投放浮标

为了尽早发现海啸，浮标必须靠近海啸发生的地方。这意味着它们必须分散地放在海底——特别是经常发生地震的洋底深沟附近。直升机将它们投放在正确的位置，然后用加重缆绳固定。

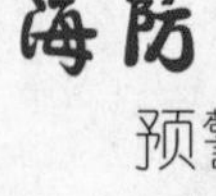

海防

预警可以帮助人们在海啸来袭前撤离沿海地区。同时，岸上自然的和人造的障碍物能够减弱海啸的威力。在日本沼津市，人们在码头入口处建造了一个巨大的门。它可以在5分钟内关闭，阻挡6米高的海啸。

严峻的未来

科学家们花了很大的精力，试图计算出下一次大地震或海啸发生的时间和地点。至今，他们最多能做到的是，在地震或海啸发生的第一时间向我们发出警报。

惊涛骇浪

别哈火山位于大西洋的加那利群岛。未来某一天，别哈火山的一部分可能突然崩塌进海里。果真如此的话，它会不会引发一场席卷大西洋的特大海啸？数百米高的巨浪可能会拍向美国波士顿、纽约和迈阿密这些城市，就像这张虚构的图片一样。所幸，大多数科学家认为这样的场景发生的可能性不大。

大西洋上的比赛

认为别哈火山可能会引发一场巨大海啸的科学家们计算出了海啸的行进速度。他们认为海啸将以720千米的时速，在火山发生崩坍1小时后淹没非洲海岸，6小时后淹没北美洲。但是其他科学家认为这样的波浪会在行进途中分解成小浪。

大型逆冲区的危险

卡斯凯迪亚“大型逆冲区”断层是一个长1000千米的海底地壳裂缝。它从温哥华岛北部一直延伸到加利福尼亚的门多西诺角。专家认为，如果该断层发生移位，那么由此引发的地震释放的能量将是圣安德烈亚斯断层移位释放出能量的30倍。断层移位可以使该地区下沉2米以上，同时摧毁西雅图等城市。

大事件时间轴

公元132年

张衡发明了第一台地动仪（图为复制品）

1906年

旧金山大地震引发了火灾，大火肆虐，很多人无家可归

1923年

日本关东大地震后，大火摧毁了东京大部分地区，死亡人数和失踪人数总计超过140,000人，360,000座房屋被毁（见下图）

公元132年

1556年

中国陕西和山西发生大地震，830,000人丧生

1775年

葡萄牙发生地震，地震引发的大海啸袭击了它的首都里斯本，海浪高达15米以上

1885年

英国地理学家约翰·米尔恩发明了近代第一台地震仪

1935年

美国人查尔斯·里克特发明了测量地震级数的里氏震级

1960年

智利发生有记录以来最大的地震——里氏震级9.4—9.6级，地震引发的海啸造成了巨大损失，连遥远的日本也未能幸免

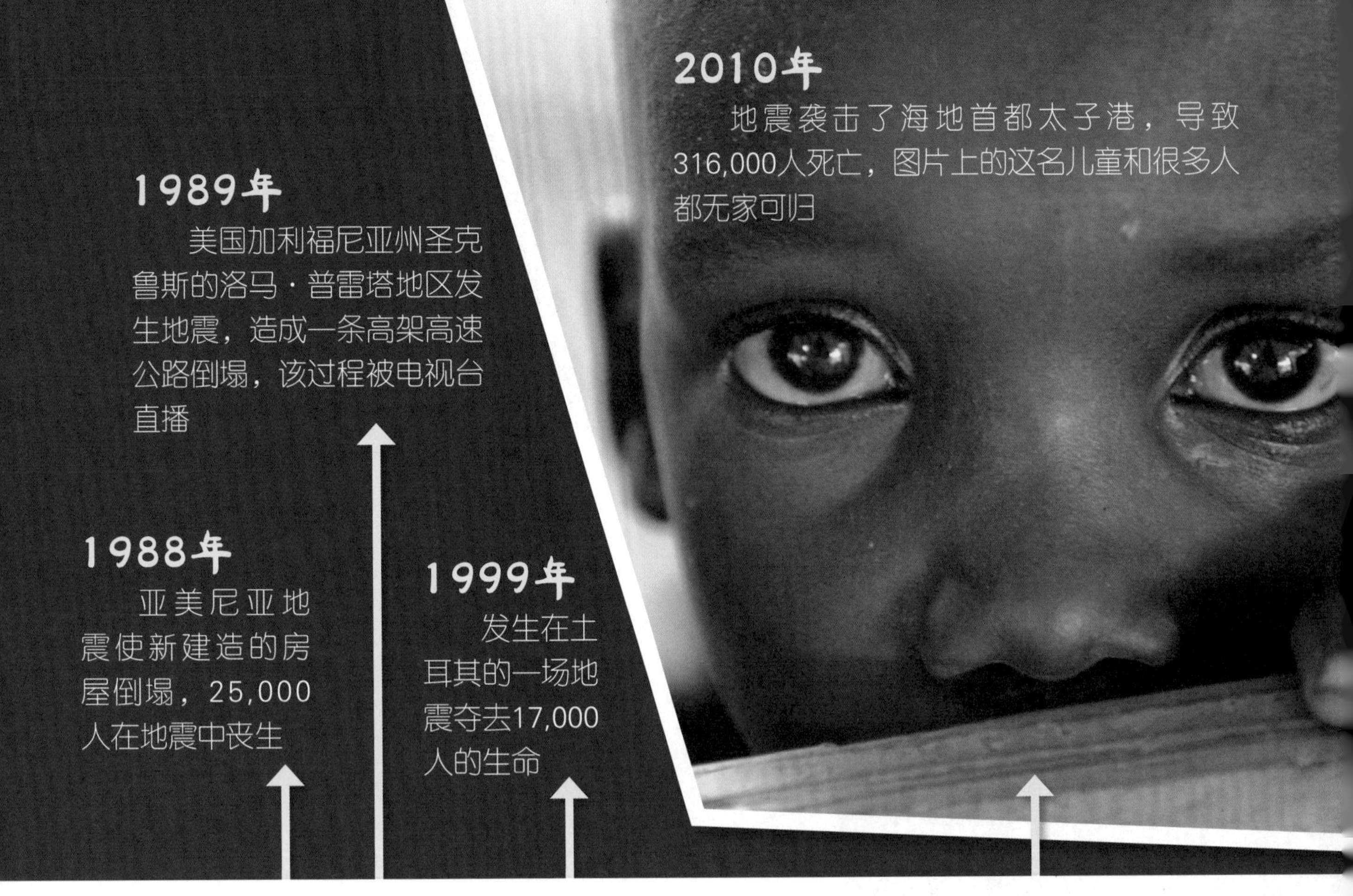

2010年

地震袭击了海地首都太子港，导致316,000人死亡，图片上的这名儿童和很多人都无家可归

1989年

美国加利福尼亚州圣克鲁斯的洛马·普雷塔地区发生地震，造成一条高架高速公路倒塌，该过程被电视台直播

1988年

亚美尼亚地震使新建造的房屋倒塌，25,000人在地震中丧生

1999年

发生在土耳其的一场地震夺去17,000人的生命

1970年

秘鲁的一场地震引发山区雪崩，造成约70,000人丧生

1976年

中国唐山遭遇7.8级地震，死亡人数达240,000人

1995年

日本神户发生7.2级地震，导致6,400人死亡

2001年

印度古吉拉特邦发生地震，造成20,000人死亡

2003年

伊朗发生地震，造成26,000人死亡，巴姆古城被毁

2004年

印度尼西亚苏门答腊岛外海的大地震引发“节礼日”海啸，造成印度洋周边地区约230,000人死亡

2005年

巴基斯坦与克什米尔地区发生7.6级地震，逾79,000人丧生

2011年

日本东海岸发生的东北部大地震引发骇人的大海啸

不可思议！

关于地震和海啸的惊人事实

失去的时间

地震能使整个地球移动！2011年日本东北部大地震改变了地球质量的分布，使地球旋转速度加快，地球上一天的时间缩短了1.6微秒。2004年苏门答腊地震使一天的时间缩短了6.8微秒。

聪明人

三名希腊科学家——瓦洛特索斯教授、亚历克索普洛斯教授和诺米考斯教授，开发了一个预报地震的系统，并以他们三人名字的首字母将其命名为“VAN”预报方法。这种方法探测在地表流动的自然电流的扰动，这种电流称为“大地电流”。他们用这一方法成功预测了一些地震的发生，但尚未有足够证据使大多数科学家相信这一方法能真正奏效。

月震

月球也会发生震动，叫做月震，而不是地震。月震通常比地震要弱。

一个小小的移动

构成地球表面的巨大岩石块叫做地壳板块，它们移动得很慢——平均每年移动距离不超过17厘米。但是，如果一个板块移动了20厘米左右的距离，就可以引发大地震。

警惕地震

有时，池塘和河道在地震前会散发出奇怪的气味。科学家认为这种气味来自地下气体的释放。此外，地面水温还可能会升高。

印度尼西亚的布罗莫火山坐落在太平洋地震带上

三维模型中的地震

近年来，地震学家已经开始将超级计算机的巨大计算能力用于预测地震。美国南加州地震中心的地震研究项目正与世界上最快的计算机之一合作，以构建一个地震发生时地下同步状况的三维电脑模型。模型会使用非常详尽的地震仪数据。

地震热点

每五个大地震中就有四个发生在环太平洋地震带周围。这是太平洋周边一个巨大的环形地带，许多地壳板块在此相接。第二大容易发生地震的地区是欧亚地震带，它跨越了土耳其、印度和巴基斯坦。

地震之最

全球死亡人数最多
中国陕西省和山西省
1556年1月23日
死亡人数：超过830,000人

近代死亡人数最多
中国唐山市
1976年7月28日
死亡人数：达240,000人

威力最大
智利瓦尔迪维亚
1960年5月22日
里氏震级：9.4—9.6级

美国史上最大
阿拉斯加中南部威廉王子湾
1964年3月27日
里氏震级：9.2级

海啸之最

全球死亡人数最多
希腊，地中海东部，克里特地震海啸
公元365年7月21日
死亡人数：可能高达500,000人

近代死亡人数最多
印度洋“节礼日”海啸
2004年12月26日
死亡人数：230,000人

浪高最高
美国利图亚湾海啸
1958年7月9日
高度：524米

速度最快
印度洋“节礼日”海啸
2004年12月26日
移动速度：800千米／小时

作者简介

约翰·法恩登，英国伦敦城市行业协会的皇家文学基金会研究员。他为成人与儿童写作了大量图书，涉及科学、技术、自然等领域，包括畅销世界的《你觉得自己聪明吗？》和《放我出去》。他六次入围英国皇家学会青少年科学图书奖，入围书目有《地球原理》《何因何果》和《人体项目》（2016）等。

Picture Credits (abbreviations: t = top; b = bottom; c = centre; l = left; r = right)
© www.shutterstock.com:
1 c, 4 c, 14 b, 19 br, 21 br, 26-27 c, 28 tr, 28 cr, 30 bl, 32 r.
11 tr = © Antonio Nardelli / Shutterstock.com, 13 tr - © Fotos593 / Shutterstock.com, 13 br = © fabiodevilla / Shutterstock.com, 29 tr = © arindambanerjee / Shutterstock, Inc, 31 tl = © CHEN WS / Shutterstock, Inc.

BC, cr = Design Pics Inc / Alamy Stock Photo. 2, l = epa european pressphoto agency b.v. / Alamy Stock Photo. 3, c = Pictura Collectus / Alamy Stock Photo. 6, c = Chronicle / Alamy Stock Photo. 7, c = Design Pics Inc / Alamy Stock Photo. 8, b = Kevin Schafer / Alamy Stock Photo. 9, cl = epa european pressphoto agency b.v. / Alamy Stock Photo. 10, c = Christian Kober 1 / Alamy Stock Photo. 11, cr = REUTERS / Alamy Stock Photo. 12, c = Bart Pro / Alamy Stock Photo. 13, b = michael seaman / Alamy Stock Photo. 13, cl = Shutterstock.com / deepspace. 15, tr = ZUMA Press, Inc. / Alamy Stock Photo. 16, c = dpa picture alliance archive / Alamy Stock Photo. 17, tl = Chrispo / Alamy Stock Photo. 18, c = Archive Image / Alamy Stock Photo. 20, c = North Wind Picture Archives / Alamy Stock Photo. 21, tl = Pictura Collectus / Alamy Stock Photo. 22, c = epa european pressphoto agency b.v. / Alamy Stock Photo. 22, b = REUTERS / Alamy Stock Photo. 23, cl = Design Pics Inc / Alamy Stock Photo. 23, br = epa european pressphoto agency b.v. / Alamy Stock Photo. 25, br = Photo Japan / Alamy Stock Photo. 25, tl = david hancock / Alamy Stock Photo. 25, tr = imageBROKER / Alamy Stock Photo. 28, bl = Granger Historical Picture Archive / Alamy Stock Ph. 28, tl = Keren Su/China Span / Alamy Stock Photo